KHOSROW M. HADIPOUR

OIL AND GAS

ARTIFICIAL FLUID LIFTING TECHNIQUES

To order additional copies of this book, contact:
Xlibris
844-714-8691
www.Xlibris.com
Orders@Xlibris.com

ISBN: Softcover   978-1-6698-0314-0
           Hardcover   978-1-6698-0313-3
           EBook   978-1-6698-0315-7

Print information available on the last page

Rev. date: 05/30/2022

CONTENTS

Published by K. M. Hadipour

Professional Petroleum Engineer

Oil and Gas Artificial Fluid-Lifting Techniques

The technical material presented below is based on forty-three years of practical experience in the oil and gas industry working for the Gulf Oil Company, Chevron USA, Pennzoil Companies, Devon Energy, and Americo Energy resources offshore and onshore in Louisiana, Mississippi, Texas, Venezuela, and New Mexico.

DISCUSSION

OIL AND GAS ARTIFICIAL LIFTING TECHNOLOGY

The purpose of artificial fluid lift is to create and apply an energy source to lift a static column of wellbore fluid out of a well to maximize and enhance oil and gas production. Hydrocarbon oil, gas, and water are originally trapped in the formation pore space under significant reservoir pressure. Some hydrocarbon oil and gas reservoirs may have high pressure, and some reservoirs may have low pressure.

When a well is perforated for production, oil, gas, salt water,

and solids will be forced out from the reservoir formation into the wellbore immediately and may flow to the surface naturally for some time (it depends upon the reservoir pressure and the fluid characteristics). Most oil-producing reservoirs will have sufficient pressure to flow for a long time. The continuous flow of fluids from the reservoir will cause oil and gas production to decline. As the well produces crude oil, water, gas, and formation solids for a long time, the reservoir pressure and energy are consumed and will no longer be able to bring the fluid to the surface.

The natural flow of oil and gas is the most cost-effective and efficient method of producing hydrocarbon oil and gas out of a well. As long as the formation pressure is greater than the hydrostatic column of fluid in the wellbore, the well may continue to flow naturally. The flow of oil and gas wells is easier to maintain and cheaper to operate. One must keep the flowing rate down to keep water, sand, and solids from entering the wellbore.

Natural well flow is based on the reservoir pore pressure and the completion techniques. Formation pressure is often referred to as pore pressure. All the oil and gas wells will have limited production recovery. The oil, gas, and water percent (%) ratios are subject to change with time.

Oil and gas wells are drilled and completed either vertically, laterally, or horizontally. Horizontal wellbores are subject to high risks and high rewards. Initial oil and gas production in horizontal wellbore drilling appears to be as high as several thousand barrels per day for several months; however, the risk and cost of borehole cleaning, troubleshooting, and artificial lifting will stay high for the remainder of the wellbore's active life.

The completion and stimulation fracturing of horizontal boreholes may consist of several comingling perforated horizons of oil, water, and gas. The drilling and borehole cleaning of horizontal wells are generally costly and risky. It may be difficult to know which zones make the most water and gas. Remedial workover and artificial fluid lifting in horizontal boreholes are always costly and challenging to face.

Reservoir Fluid Behavior

Understanding reservoir and fluid characteristics can be valuable in artificial lifting decisions.

There are basically three types of reservoirs:

- Water drive reservoir
- Solution gas drive reservoir
- Gas cap expansion drive reservoir

(Misunderstanding the above reservoir characteristics before selecting artificial lifting equipment could cause major cost problems later.)

When the reservoir pressure becomes equal to the hydrostatic column of fluid in the wellbore, the well will stop flowing naturally. The fluid level in the wellbore will stall at or near the surface, with some gas bubbles in the solution. The bubble point is calculated to understand the reservoir characteristics. The bubble point is the function of the reservoir temperature, gas/oil ratio, and oil and gas gravity.

When the reservoir pressure declines below the hydrostatic head of the column of liquid in the wellbore, fluid will stall below the surface inside the casing/tubing string. The rapid decline of reservoir pressure will cause the hydrostatic head to become equal to or greater than the reservoir pressure. The fluid in the wellbore predominantly turns to heavy salt water.

The hydrostatic head is defined as pressure exerted by the column of fluids inside the tubing and/or casing string (the fluid is referred to as oil, water, and gas in the oil field). When a well stops flowing oil and gas, naturally, it will be reported as a *dead well*.

An artificial lift begins when an oil and gas well ceases to flow naturally because of the depletion of reservoir pressure and natural fluid flow. Stimulating and fracturing the reservoir may extend the natural fluid flow longer. Once an oil or gas well stops flowing fluid, the artificial lift may begin with basic fluid swabbing to prolong the natural flow of the oil or gas well as long as possible while planning and preparing for an appropriate artificial lifting system.

Oil and gas reservoirs contain pressure. Some reservoirs may contain high pressure, and others may have low pressure (depending on the reservoir formation characteristics). Artificial lifting will reduce the hydrostatic head of the column of fluid significantly.

$$\text{Hydrostatic pressure} = 0.05195 \times \text{fluid density} \times \text{fluid height}$$

Hint: One gallon of fresh water is equal to one pound and is equal to 231 cubic inches.

Do You Know What Swabbing Is?

Swabbing is a temporary artificial lifting technique of pulling water, oil, and gas from a well using a swab unit (rig), steel wire rope, and mechanical swabbing tools. Swabbing the liquid out of the tubing string will reduce the hydrostatic head, and it may help the well flow naturally.

Mechanical swab tools and equipment consist of the following:

- Swab unit mast (swab rig)
- Swab lubricator with pressure control pack-off elements
- Swab line (9/16" steel wire rope)
- Rope socket
- Steel swab bar/bars
- Mechanical spank jars

- Swab mandrels
- Rubber and/or steel swab cups

Effective swabbing will begin when the reservoir's bottom-hole pressure depletes or declines to a level lower than the hydrostatic column of liquid in the tubing and casing, preventing the well from flowing naturally (check the perforations to make sure they are not covered with sand).

The swabbing results may be utilized to determine the total daily fluid entry into a well (reservoir productivity and deliverability). Do not swab the formation sand and solids into the wellbore. An accurate reservoir output will assist you on the artificial lifting design and the selection of appropriate artificial lifting tools and equipment.

An effective swabbing method must be conducted in conjunction with a good-standing mechanical isolation packer and a seating nipple on the tubing string that is set above the perforations. Through the tubing, swabbing must be conducted with an isolation packer on the tubing string (set the packer at one hundred feet above the open perforations to avoid pulling formation sand in the hole).

When the well starts flowing, reduce the back pressure at the production facilities to help the well unload and flow for as long as possible. When swabbing through the production tubing, the tubing string must be in good standing without any hole/holes (never swab a well down to the seating nipple). A standard API seating nipple must be installed above the isolation packer before swabbing (make sure the no-go on the swab mandrel

will not go through the seating nipple).

There are two distinct swabbing methods:

- Through tubing
- Through casing

The daily production rate and the volume of fluid from a reservoir must be known before an artificial lifting design is applied (read *The Field of Swabbing* by the author of this book). If swabbing a well is no longer effective to keep an oil or gas well flowing naturally, an artificial lifting method may be implemented to change the wellbore fluid from static to kinetic energy lift.

The rapid pressure depletion of a reservoir is due to poor completion techniques and the waste of reservoir gas into the atmosphere (blowing down and flair burning). More than hundreds and thousands of pumping units appear to be on time clocks across oil fields (nearly 85% of oil wells are on time clocks).

Artificial Lift Methods in the Oil and Gas Fields.

Artificial lift is a vast subject to cover or discuss in this book. However, I will discuss the most important and practical parts of all the artificial operating techniques as we go along through this chapter.

- Keeping a well flowing naturally is highly cost effective and easy to operate.

An artificial lifting operation is a costly method, no matter how you may look at it. The choice and selection of an artificial lifting method is at your discretion. Everyone is in a hurry in the oil field to get that barrel of oil and gas out of the hole. I have seen many costly mistakes in drilling and the completion and selection of artificial lifting methods by new engineers and managers in particular.

- Evaluate the artificial types and decide what is good for the well (not what you like).

Evaluate the fluid characteristic subject to artificial lifting. Select the best alternative lifting method that is effective and efficient with a longer lifetime. Consider the effects of the fluid flow from the reservoir into the subsurface equipment. Understand the desirable and undesirable characteristics of the reservoir fluid, such as the following:

- Well depth (open perforated depth)
- Well condition(casing mechanical integrity)
- Volume of fluid per day
- Rate
- Temperature
- Gas pressure
- Gas bubbles

- Gas/oil ratio
- Oil viscosity
- Fluid density
- Mud/sand/sludge
- Corrosion type
- Emulsion
- Turbulence and fluid surge
- Fluid heading (natural or artificial)

Practical Artificial Lifting Methods in the Oil and Gas Fields

There are basically two distinct methods of artificial fluid lift:

A) The mechanical artificial lifting method
B) The artificial gas lifting method

A. The Mechanical Artificial Lifting Method

1. Artificial lift using swabbing tools and equipment (short-term lifting)
2. Artificial lift using sucker rods and a beam pumping unit

(using a pump jack)
3. Artificial lift using hydraulic or jet pumps
4. Artificial lift using electric submergible/submersible pumps
5. Artificial lift using a rotary pump/progressive cavity pump (PCP)
6. Artificial lift using plunger lift tools (utilizing reservoir or compressed gas pressure)

B. ARTIFICIAL GAS LIFTING METHOD

The gas lift method is not mechanical fluid lifting. On the artificial lifting method, a high level of dry natural gas pressure is injected through the casing annulus to lift the wellbore fluid.

The gas lift method is the most effective and efficient method of all artificial lifting technology (it requires sufficient natural gas pressure and gas volume). Forty thousand barrels of fluid per day can be lifted using a gas lift system.

Artificial lifting is costly, and the mistakes of operating artificial lifts are more costly because of the selection of incorrect artificial lifting types, lifting designs, and operating techniques as well as poor wellbore managing.

The choice of method is at your discretion; go for it if you know what you are doing.

I will explain the mechanics on each of the above artificial lifting methods below.

Approximately 90% of oil wells in the United States are on some sort of artificial lifting method. Artificial lifting requires planning, design, knowledge, and practical experience. The components of an artificial lift are similar to the pieces of a puzzle that must be engineered and put together with accurate design and balance from start to finish.

Some of the artificial lifting methods consist of many moving components to operate and may be subject to repeated production equipment failure.

Information for Surface and Subsurface Equipment Designs:

- Reservoir input (fluid drawdown; fluid production per day)
- Accurate calculation and selection of appropriate lifting equipment
- Lifting depth (the subject of an accurate study)
- Selection of subsurface equipment types and sizes
- Casing and tubing mechanical evaluation (tubing, casing, and downhole integrity)
- Wellbore fluid evaluation (the volume of oil, water, and gas subject to lift)
- Selection of subsurface fluid pumping depth (packer and installation depth)
- Selection of tubing size and grades for a specific wellbore depth (H-40, J-55, N80, L80)

- Wellbore temperature, pressure, fluid density, corrosion evaluation, and **cleanness of fluid subject to artificial lifting**
- Availability of supporting elements/resources and equipment (natural gas and power)
- Selection of surface supporting equipment
- Selection of sucker rod string (one-size rod string vs. multiple tapered rod string)
- Cost and economic impact (production versus payoff)
- Available resources to accurately engineer the artificial lifting
- Bottom hole pressure
- Bottom hole temperature
- Fluid gradient
- Static fluid level

Artificial lift designs are based on the following information:

- Well depth — The plug back total depth (PBTD)
- Perforation depth — The depth of perforated holes in the casing at the zone of interest

- Bottom hole temperature (BHT) — The temperature at the midpoint of open perforations
- Static fluid level — The well fluid level at the static condition
- Static bottom hole pressure — The survey of wellbore pressure at the static condition
- Productivity index (PI) — The ratio of wellbore production to wellbore bottom hole flowing pressure
- Fluid density — The fluid weight (fresh water = 8.33 pounds per gallon vs. salt water = 9.4 pounds per gallon)
- Fluid viscosity
- Tubing size — The outside diameter of the tubing (2", 2 ⅜", 2 ⅞", 3 ½", 4.5")
- Casing size — The outside diameter of the casing (13 ⅜", 9 ⅝", 8 ⅝", 7", 5", 5 ½", 4 ½")
- Anticipated well production (based on the actual daily well test or fluid swab test)
- Anticipated theoretical daily production versus actual daily production (oil, water, and gas)
- Oil/water ratio
- Kickoff pressure/operating pressure
- Available gas lift/gas pressure
- Statues of sand, mud, and solids (how clean the well fluid is going to be)
- Gas/liquid ratio — The ratio of gas divided by total liquid production volume
- H2S/CO2 wellbore environment
- Wellbore mechanical condition — The physical condition of a casing and wellbore
- Other important downhole information (e.g., gravel pack, sand, shale, mud, solids)

Artificial lifting equipment will not operate properly in any oil or gas wells that produce sand and shale. The equipment will be subject to failure or become stuck. Sanded-up ESP equipment may cost you $200/foot to fish and to clean up the wellbore.

Be careful in your artificial lifting designs. Stop changing from one artificial lifting method to another (it will cost you). The cost of changing equipment in any artificial lifting well is high (e.g., changing submergible artificial lifting to gas lifting or beam pumping).

Carefully evaluate your initial lifting design/designs (get a second opinion if needed). Too many mistakes and poor decisions in artificial lifting designs may break down the financial foundation of any small company (e.g., dry drilling, bad completions, and poor lifting designs).

Many tools and equipment lying down in oil fields are going to waste as the result of changing from one piece of artificial lifting equipment to another without any preplanning (you are costing the company and the shareholders a great deal). Risky artificial lifting ideas will make well productivity less attractive.

The Artificial Lift Methods in Oil Fields

The Fundamental Principles of Artificial Lift using Pump Jacks

Nearly 85% of oil wells in the United States are operating on the sucker rod–beam pumping method (sucker rods and pump jacks).

Beam pumping consists of two major parts:

A) The surface equipment (beam pumping unit)
B) The subsurface equipment (subsurface fluid pump and rod string)

The surface pumping unit and its related components consist of the following:

1. The pump jack — The surface beam pumping units appear in different sizes and shapes and may be referred to as pumping units or pump jacks.
2. The prime mover assembly — The prime mover is the main source of energy, which consists of the electric motor and/or the gas-driven engine.

3. The commercial electrical power consumption (electric provider)

Subsurface equipment consists of three distinct major parts:

1. The subsurface tubing string, of various sizes (ranging from 2 ⅜″ to 4 ½″)
2. The subsurface fluid pumps, of various types and sizes
    a) Insert fluid pumps
    b) Tubing pumps

3. The subsurface sucker rods of various sizes, types, and grades
   a) Steel sucker rods
   b) Fiber sucker rods

CHAPTER

THE PUMPING UNITS AND RELATED EQUIPMENT

- The pumping units (pump jacks)
- The prime mover (electric motors or gas engines)
- The electric power source or natural gas-driven engines

A beam pumping unit is basically a large piece of steel equipment that may be called the pumping unit, workhorse, pump jack, or grasshopper unit. The main function of a beam pump is to reciprocate the sucker rods and the subsurface fluid pump by lifting the sucker rods and the production fluid.

The pumping unit (beam pump) is driven by an electric or gas-operated prime mover as the energy-generating source to make the pumping unit to operate (stroke up and down) and reciprocate the sucker rods, and subsurface fluid pump components to displace wellbore fluid to surface)

The prime mover is basically an alternating current (AC) electric motor of various levels of horse power and/or a gas-driven internal combustion (IC) engine (such as natural gas or propane gas). The prime mover is directly linked to the power conversion system.

Do You Know How a Pumping Unit Works?

The pumping unit consists of several moving components. Stay alert and stay safe. Working on pumping units may cause serious bodily injury or death. Most surface pumping units are mechanical operating equipment in oil fields. All the components of a mechanical pumping unit are linked and/or connected together (too many moving components).

The generated power energy from the prime mover transmits force to the pumping unit components above the ground and transfers the energy down to the subsurface sucker rods and the fluid pump down the well. The basic function of a pumping unit is to lift and reciprocate the downhole rods by changing the rotary motion at the motor to kinetic energy. The prime mover and the counterbalance stored energy are the main force in lifting the sucker rods and fluid loads (dynamic load consists of all the subsurface weights)

The Sequence of Motions of Beam Pump Components

- Generated Electric Power (Power Plant)

The electric power transmits energy (electromagnetic) to the prime mover on the pumping unit, making the motor's shaft rotate at the required horse power (revolution per minute or RPM).

A prime mover can be any generating power source—an electric motor, a steam engine, or an IC gas-operating engine using propane gas or dray natural gas

- Prime Mover (Electric Motor)

The prime mover is directly connected to power conversion. The prime mover delivers the generated power by rotating two pulleys on the tail of the pumping unit (a small sheave at the prime mover and a large sheave or flywheel on the pumping unit's gear box

or gear reducer).

- The pulleys are linked together with V-shaped rubber belts that extend from the small diameter pulley on the prime mover to the large diameter pulley (flywheel) on the pumping unit's gear box shaft.
- The power band rubber belt is the main link of energy from the prime mover to the large flywheel pulley attached to the gear box shaft (on the pumping unit).
- Changing the size of the sheave on the prime mover will increase or decrease the speed (RPM) on the pumping unit.

- The shaft on the prime mover operates with a rotary motion.
- Changing the sheave on the prime mover to increase or slow down the RPM is basically cheaper and easier than changing the large flywheel on the gear box.

- When the attached pulleys rotate (rotary motion), the gear box shaft on the pumping unit will turn, and so do any attached components to the gear box (The seesaw action of the pumping unit is based on counterbalance).
- The gear box torque power will force the attached heavy crank arms

to rotate with a good counterbalance effect. Torque is the applied force to rotate the system. The shaft on the prime mover usually operates at high RPM speed but at lower torque.
- The downhole sucker rod's weight is the major part of counterbalancing the surface beam pumping unit. On the down stroke,

the rod string will fall by its own weight.

- The power generated by the prime mover and the counterbalancing force are used to reciprocate the rods.
- You cannot counterbalance a beam pumping unit without the weight of the downhole sucker rod string (just like a weight scale balance).
- The cranks transmit energy to the wrist bearings and the crank arms in a rotating motion, raising and lowering the arms (the arms are attached to crank bearings).
- The rotation and revolution of the crank arms on the unit will cause the equalizer bearings to force the walking beam and the attached horse head to go up and down (bowing or nodding motions).
- The nodding motions of the horse head will force the attached wire bridle and the connected carrier bar to lift the attached polished rod, the sucker rod string, and the subsurface fluid pump to stroke and displace fluid in the tubing string on the upward motion.

See the attached pictures of the various sizes and shapes of pumping units. Caution: Some pumping unit cranks are manufactured to rotate clockwise and/or counterclockwise.

- Some pumping units are designed to rotate counterclockwise only.

See the stamped arrow on the cranks for the correct rotations (check the unit before wiring). Look at the Lufkin Mark II

and IV pumping units for the correct rotations. Based on the rod string weight, the beam pumping unit can be balanced with or without counterbalance weight segments for the prime mover to function properly (the counterbalance weights may be relocated and/or removed to obtain correct counterbalancing).

- The direction of generated force through the beam unit is basically from the prime mover to the cranks and pitman arms, up the equalizer bearing and onto the walking beam.

The walking beam supports and delivers the force onto the horse head and down the rod string.

The gear box (gear reducer) is the center of all the actions.

- The gear box (gear train) is designed to reduce speed (RPM).

To protect the gear box, the oil level must be kept at the recommended level with suitable viscosity and free of sludge, water, and shavings.

Adverse operating conditions can cause gear box damage:

➢ Intermittent operation (on/off and time clocks) on the pumping unit (this will cause wear and damage to the gear reducer)
➢ Excessive sand or dust particles and sludge in the gear box
➢ Corrosive sour gas fume conditions and water mixture

When the pumping unit mechanically strokes up and down, it will cause the subsurface fluid pump and the sucker rod string

to reciprocate, create suction, and draw and displace fluid (oil, water, and gas) on the upward motion.

The up and down motions of the pump jack will cause the subsurface pump to reciprocate, making the valves function by opening and closing, thus drawing the well fluid into the pump's chamber and displacing oil, water, and gas on the upward motion.

The Purpose of the Pumping Unit's Bridle

The bridle on the horse head (**drive head**) is one of the necessary components of the beam pumping unit. The bridle is made from flexible steel wire rope and/or flat belts driven on some beam pumps such as Rotaflex units.

The bridle wire rope has two major functions:

- Pull the rods with the dynamic load on the up stroke
- Hold the rod string from dropping into the well

The bridle or the horse head (drive head) does not have the power to push the rods down the hole. With proper counterbalancing, the rod string will fall by its own weight, and the weight of the free-falling sucker rods will force the horse head (drive head) to move on the down-stroke motion faster. The falling rod string will normally store energy in the counterbalance and helps on the upstroke motions (see the counterbalance weights on the pumping units).

Counterbalance will not be effective if the bridle moves faster than the sucker rods on the down-stroke motions. If the rod string falls too slow on the down stroke and the horse head or bridle drops too fast, the polished rod clamp may lose its momentum and connecting space with the carrier bar. This will cause a large space between the polished rod clamp and the carrier bar and will make the carrier bar jar up onto the polished rod clamp on the upstroke motions. This abnormal action may cause kinks or throw the bridle off the horse head.

Install a bridle guard on the horse head to prevent the cable from jumping off the head

A slow-falling sucker rod string is due to the sticking of a pump plunger (and downhole obstructions or may be due to shallow parted rods.

Undersized stuffing box packing elements, broken rod guides, a sanded-up pump plunger, parted rods at shallow depth, paraffin buildup, and/or parted tubing string

could cause the bridle to be thrown off the horse head on the down-stroke motions. (Lubricate and check the wire rope for kinks and broken strands regularly.)

The Beam Pumping Units

Pump jacks are valuable equipment and are the best artificial lifting selection for low fluids, remote oil, and gas wells. The proper use of pump jacks is often misunderstood among some people in the oil field. They like to reciprocate the pumping units too fast.

There are several types, shapes, and sizes of pumping units in the oil field:

- Vertical flat belt-driven pumping units (Rotaflex) with long strokes

- Air-balance units with long strokes

- Conventional crank-balanced pumping units
- Counterweight beam pumps (the weights are set at the tail of the walking beam)
- Hydraulic beam pumping (using a hydraulic cylinder)
- The Lufkin Mark–type pumping unit

Each of the above pumping jacks may have significant advantages and disadvantages in operation and maintenance (one needs to see and learn by experience). The basic principle of all surface and subsurface pumps are the same. Some are more efficient and better designed than others. ***Choose a pumping unit that serves the purpose. Choose quality-over-quantity products.***

The Advantage of Lufkin Air-Balance Pumping Units

- Offer slow and long strokes, up to 20′ in length
- No counterbalance weights needed
- Lower installation cost
- Automatic counterbalance options using air pressure
- Safety shutoff system
- Shorter and lighter unit

The air-balanced walking beam assembly uses high-pressure air that is compressed in a cylinder to balance out the load. The air cylinder is one of the components of the air balance unit

and hinged in the front of the unit. The high-pressure air cylinder is filled with high-pressure air supplied by a compressor. Air is usually compressed between the piston and the cylinder on the down-stroke motion. On the polished rod upstroke motion, the pressure of the air helps lift the walking beam and the sucker rod load.
The air pressure inside the cylinder is at its highest at the beginning of the up stroke and is at its lowest at the beginning of the down-stroke motion. When the weight exerts more force, the air is compressed. The compressed air in the cylinder will store energy. The compression of air will increase the pressure and the amount of force to lift.

To balance the rod load, the pressure of the air in the cylinder must be high enough to lift the weight of the rod string plus the fluid load. If the air cylinder loses air pressure, the cylinder can be refilled using an auxiliary compressor. Excess air can be vented through a safety valve. For the purpose of effective counterbalancing, the air balance unit needs a larger volume of air. Air balance units normally run smoother than the counterweight units because of air cushioning. The system can be adjusted while the unit is operating.

The Advantage of Rotaflex Beam Pumps

Long-Stroke Pumping Unit

A Rotaflex beam pumping unit

may be used on a shallow and/or deep wellbore ranging from two thousand feet to ten thousand feet. A Rotaflex beam pumping unit will also offer a greater advantage in stroke length, with slow strokes per minute. The advantage of long and slow strokes in beam pumping will reduce wear and tear on the sucker rods and the subsurface pump and will allow the fluid to properly fill up the pump chamber, preventing a water hammer.

Slow strokes may prevent over-travel and under-travel motions. Rotaflex units will have all other options including but not limited to variable speed drives (VSDs) and monitoring systems.

There are several sizes and models of Rotaflex units in the market to select based on the wellbore requirements. These units are unique and efficient on several selected options to increase production and reduce wellbore repairs.

a) Designed to clear space for workover rigs using integrated hydraulic systems to displace and move the unit back twelve feet from the wellhead to safely rig up on the well
b) May offer incredible stroke lengths of nearly thirty feet
c) Major access to repair key components
d) Designed with a hydraulic break system
e) Rod rotator system
f) VSD computer system (variable speed drive)
g) Good counterbalance options
h) Less moving parts at the surface

Artificial lift beam pumping units have been well known to the oil industry since the early 1900s. Pumping unit manufacturing companies such as Lufkin, Oil Well, Bethlehem, American, EMCO, and others are the early manufacturing companies of pump jacks.

The early pumping units were made of simple wooden beams with gas engines or steam engines and electric prime movers. The units were operated by steel rod liners laid parallel to the ground surface and connected to different wells. The rods were laid down flat on the surface and extended from a well to a main powerhouse under a shed for protection (e.g., early pumping wells in Batson, Texas). Modern beam pumping units are manufactured in various physical shapes, sizes, and stroke lengths and are selected based on wellbore depth and fluid pumping capability applications (new pump jacks are built to be stronger, safer, and faster).

Across the oil-rich fields of West Texas, the land is covered with the abovementioned pump jacks as far as one can see (Monahan, Odessa, Midland, Big Spring, the Loco Hills of New Mexico, and Denver City, Texas, to mention a few). Many more pumping units will be installed across the oil fields with the expansion of drilling operations and demand for more oil and gas production across the world in the years to come.

Applications of Pumping Units (Pump Jacks)

Surface pumping units or pump jacks are designed based on specific wellbore pumping depth (shallow wells require smaller units, and deeper wells require larger and stronger pump jacks to work with). Pumping units must generate sufficient power to safely lift the entire rod string with the total dynamic fluid load (rod string, water, oil, gas, and solids). There are many shapes and sizes of pumping units available in the oil field

Pumping Unit Identification

The identification of a pumping unit clearly indicates the pumping unit's basic parameters—pumping unit model, loading range, and stroke lengths—so that you may select the correct-sized pumping unit for specific wellbore depth. The pumping unit identification is normally attached on the face of the gear box (gear reducer) and often appears on the part of the Samson post.

Example:

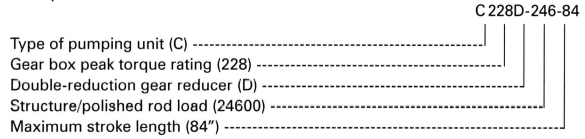

C 228D-246-84

Type of pumping unit (C) ---
Gear box peak torque rating (228) ----------------------------------
Double-reduction gear reducer (D) ----------------------------------
Structure/polished rod load (24600) -----------------------------------
Maximum stroke length (84") -----------------------------------

Type of pumping unit:

A = Air-balanced unit
B = Beam-balanced unit
C = Conventional pumping units
D = Double gear reducer

The beam balanced units are the smaller units for the shallow wells: B-25, B-40, B-50, B-57, B-80, and B-114. Small, mid-range, and large units are available to fit specific wellbore depths ranging from one thousand feet to twelve thousand feet:

C57-D	88" to 144" stroke
C80-D	42" to 54" stroke
C114	-D48" to 84" stroke
C160-D	54" to 75" stroke
C228-D	64" to 86" stroke
C320-D	88" to 144" stroke
C456-D	120" to 168" stroke
C640-D	120" to 194" stroke
C912-D	144" to 210" stroke
C1280D	168" to 300" stroke
C1824-D	192" to 300" stroke
C2560-D	240" to 300" stroke
C3648-D	240" to 300" stroke

The Components of a Beam Pumping Unit (Pump Jack)

The components of a beam pumping unit may vary depending on the unit type:

- Carrier bar
- Wireline (bridle)
- Horse head
- Walking beam
- Equalizer bearing
- Saddle bearing (Samson bearing)
- Samson post
- Safety ladder
- Base beam (I-beam frame)
- Brakes

- Brake lever
- Brake cable
- Brake drum
- Prime mover
- V-belt drive
- Pitman
- Pitman arms
- Wrist pin bearing (crank pin bearings)
- Cranks
- Crank shaft
- Gear box (gear reducer)
- Weights (counterbalance weights)

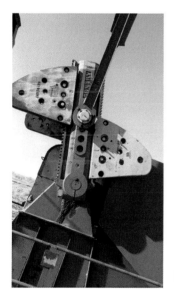 

- Belt guard (belt cover)

Most of the components of the beam pumping units are made from carbon steel (I-beam, angle iron, or cast iron).

Caution: Pumping units consists of many moving parts. Stay alert and stay clear from pumping units when they are in motion. Serious injury or death may occur while working on pumping unit components.

Components of a Pumping Unit

1. Base Beam — Made with a large heavy-duty steel I-beam frame to support the weight of all the components of the pumping unit above the base.

2. Samson Post — Engineered to support the bearings, the walking beam, and related parts (the Samson post may be designed differently on some pumping units with basically the same principle).

3. Walking Beam — A large heavy-duty carbon steel I-beam above the Samson post mounted on the center bearings (it is designed to hold the horse head). The main function of the walking beam is to raise and lower the polished rod assembly. The walking beam supports the horse head loads and lifts the subsurface rods. The walking beam obtains power from the pitman arms by rocking on a pivot on the Samson post (saddle bearing). The nodding motion of the walking beam is caused by the full rotation of the pitman arms.

4. Tail Bearing — The "equalizer" bearing, which assists the walking beam's up and down motions. The equalizer bearing carries loads similar to the center bearing and must be packed with lubricating grease periodically. The saddle bearing and tail bearing are greased through a grease fitting that extends up to the bearing housing. On some pumping units, the equalizer bearing is fastened to the walking beam near the horse head. Changing the distance between the equalizer bearing and the Samson post may change the length of the polished rod stroke.

5. Saddle Bearing — The "center" bearing, which supports and assists the walking beam's motion.

6. Horse Head (Drive Head) — A welded curved heavy-duty carbon steel metal head to support and hold the hanger assembly (bridle and carrier bar) while reciprocating the sucker rod string with the fluid load. Most horse heads are designed with curved faces.

7. Bridle/"Wire Rope" — I consider the wire rope as basically a machine with moving strands. The steel wire rope loops at the back of the horse head and extends down to the face of the horse head to hold the carrier bar and heavy sucker rod loads below. (Strands on the cable require lubrication regularly.) The wire rope/bridle is designed to pull the dynamic load with an up-string motion. The sucker rod weight will pull the head and the bridle on a down-stroke motion. The bridle and carrier bar must be hung straight and level. The pumping unit bridle may be designed differently on some styles of pumping units.

8. Carrier Bar — A specially made device. The purpose of the carrier bar is to hold the wire bridle and the polished rod clamp with the entire dynamic load while in motion and/or stationary/at rest. On the reciprocation motions, the entire dynamic load of the rod string will be transferred to the carrier bar through the polished rod clamp. It is important that the carrier bar be hung flat, straight, and level. If the carrier bar is not level, the polished rod clamp may force the polished rod to a bending angle and break below the polished rod clamp, causing the entire rod string to drop into the well. The main purpose of the bridle and carrier bar is to lift the rods on the up stroke and hold the rod string on the down stroke.

9. Gear Box (Gear Reducer) — As the name implies, the gear train may be called the gear reducer and is designed to reduce speed (RPM). Big pumping units must be operated on a long stroke and operate for not more than six or seven strokes per minute. The gear box is the main engine to deliver rotary motion to the reciprocation motion of the lifting sucker rods. Double-reduction gears trains are used on most pumping unit gear boxes.

The lubricating oil in the gear box must be checked and kept clean to prolong the life of the gear box. Special seals on the gear box shaft must be checked for leaks (check the oil level in the gear box every six months). The gear box oil must be of a certain viscosity as recommended by the pump manufacturing company. Sudden heavy load on the gear box will cause gear damage (may break the teeth on the sprockets). The gear box oil must be free of water, sludge, and steel shavings and must be checked once every year to make certain that the oil level is sufficient and kept clean. Leaking oil from the packing and drive shaft must be noted daily.

The driven pulley on the V-belt drive is normally attached to the smallest gear in the reducing gear. The load on the pumping unit gear box should not exceed the peak torque in the gear box. Limited load must be applied to the pumping unit's gear box.

The power conversion on the pumping unit consists of the following:

- Main power supply to the pumping unit (the prime mover)

- V-belt drive
- Gear reducer
- Crank shaft and cranks

The rotary motion at the prime mover is reduced in speed before it is changed to reciprocating motion.

Counterbalance Weights

Each pumping unit is designed with special removable counterbalance weight blocks. The counterbalance weights are made of heavy cast iron and designed to fit a particular model of pumping unit only. The counterbalancing on the pumping unit uses energy from the falling rod string. Counterbalancing stores energy on the down stroke to lift the sucker rod load on the up stroke. With good counterbalance, the prime mover uses less power to lift the rods.

There are two types of counterbalance weights:

- The beam counterbalance weights are installed on the pump unit's I-beam at the tail of the unit.
- The crank counterbalance weights are installed on the crank arms of the pumping units (some counterbalance weights may be hollow inside).

The counterbalance weights cannot be adjusted when the unit is in operation (stay clear; stay safe). Smaller units will have weights on the tail of the walking beam, and larger units will have the weights bolted on the cranks. The weight can be installed or removed to balance the pumping unit. Weights are auxiliary equipment for counterbalancing only.

10. Cranks — Used to hold the counterbalance weights and the pitman arms. Cranks rotate the arms in cycle motions. Cranks are made of heavy cast iron and often can be used as counterbalance weights alone. Some smaller pumping units may have the counterbalance weights on the tail of the walking beam. The counterbalance weights are subject to change or removed if necessary to balance the unit (stay clear from moving cranks.)

11. Wrist Pins — It is very important to check and maintain the wrist pins and wrist pin bearings so that they remain in good operating condition:

- Clean the wrist pin and wrist pin hole thoroughly with solvents when needed.
- Dry the shaft and hole completely.
- The pin and holes must be free of grease, grit, paint, rust, and burr before installation.
- Never hammer the bearing housing during installation.
- Thrust the wrist pin into the hole, and install the wrist pin nut by hand first and then tighten the nut fully using a sledge hammer on the wrist pin.
- Install a cotter pin to lock up the system.
- Check and re-tighten the wrist pin nut after twenty-four hours of operation.

Caution: Pumping units consist of many moving parts. Stay alert and stay clear from the pumping units when they are in motion. Serious injury or death may occur while changing the pumping unit components (keep guard around the unit).

12. Pitman Arms — Used to activate the equalizer bearings and rotate the cranks in a circular motion to lift and lower the walking beam (nodding motion). The transmitted power from the conversion to the walking beam is provided by the pitman arms.

KHOSROW M. HADIPOUR

Safety Color Codes on the Pumping Unit

There are several spots on the pumping unit's moving components that must be painted in bright visible orange or red for safety purposes:

- Tip of the horse head
- Tips of the carrier bars
- Tips of the counterbalance weights
- Tips of the pitman arms
- Wrist pins

13. Counterbalancing — Done by using special shaped weights designed to fit and balance a pumping unit's motions. Counterbalance weights are engineered to fit special models of pumping units and may appear at the tail of the beam pump or bolted on the crank pitman. When you hang the rods on the beam, the rod string dynamic load must be balanced with the counterbalance weights on the unit (similar to scale balance). Too much rod weight is referred to as **"rod heavy"** (the horse head stays down to the wellhead). Too much counterbalance weight is referred to as **"weight heavy"** (the horse head stays up high). You must balance the weights to obtain smooth pumping counterbalancing. Counterbalancing will save energy and prevent rapid swings of the walking beam.

Supporting Tools and Equipment That Make the Beam Pump Work

14. V-Belts — The V-belts have a V-shape design at the bottom to accommodate the sheave groove. V-belts connect the prime mover sheave and the gear box sheave to transmit power. V-belts are designed with either single rubber belts or power band V-belts. Sheaves are actually called pulleys with various sizes and numbers of grooves. As power is transmitted through the V-belt and the gear box, the high speed will be changed to high torque at the gear shaft.

Several V-belts can transmit more power than one belt (the principle of horse power and carriage). The power band V-belt is stronger and capable of transmitting power more efficiently than single V-belts. The V-belt must be adjusted with proper tension to transmit power efficiently. Loose V-belts will waste power. They slip and stop the pumping unit from proper reciprocation. When one of the several belts in a set becomes worn, it is a good practice to change the entire set. Check the pulleys for wear and sharp cuts.

The entire belt drive must be protected with a belt guard for safety reasons. Too-tight belts will stretch and waste power. Tight belts may burn or break down. In any

V-belt drive, the belt should give by being pressed down from the top based on the thickness of the belt (a 1" thick V-belt should give 1" slack for every 4" of length).

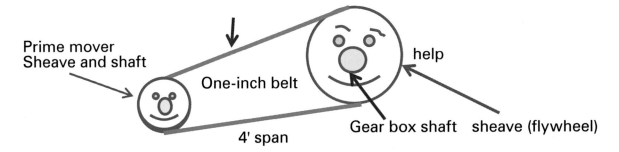

Prime mover Sheave and shaft

One-inch belt

help

Gear box shaft   sheave (flywheel)

4' span

Belts may stretch and break because of improper application, stuck sucker rods, bad counterbalance, or shallow parted sucker rods. Keep the sheaves and belts dry.

$$\text{Belt Length} = 2C + 1.57\,(D + d) + \frac{(D-d)2}{4C}$$

D- Large sheave

$$D = \frac{RPM \times C}{SPM \times R}$$

d- small sheave

$$d = \frac{SPM \times R \times D}{RPM}$$

C- shaft center distance

$$SPM = \frac{RPM \times d}{R \times D}$$

$$RPM = \frac{SPM \times R \times D}{d}$$

R- Gear ratio

$$SPM = \frac{RPM \times d}{SPM \times D}$$

❖

15. Prime Mover — An electric motor or IC gas engine. Install the correct-sized prime mover to start the operation correctly. The electric motor must generate enough horse power (HP) to lift the cranks with the entire dynamic load.

$$HP = \frac{BPD \times pump\ depth}{45,000}$$

The dynamic load consists of the rod string, the fluid inside the tubing string, and friction. The size of the prime mover depends on the size of the pumping unit, the well depth, the rod string configuration, and daily production. An undersized prime mover will burn and will not lift the cranks. Underbalanced or overbalanced pumping units will blow electric fuses and may burn the electric motor, draw more electricity, or burn the V-belts. Stay clear while working around counterbalance weights.

If the pumping unit is not reciprocating properly, the reason could be one of the following:

- Counterbalance problems
- Downhole problems (stuck rods or shallow parted rods)

- Bad order fuses or wires inside the panel (single phasing)
- Power source (bad order transformer)
- Bad electric motor (undersized or single phasing)
- Unit is weight heavy (unbalanced unit)
- Unit is rod heavy (unbalanced unit)
- Gear box problems (cranks will not move)
- Wire connection problems
- Power bands that are too loose
- Brakes that are too tight

Beam Pumping Unit Installation

Before you install a pumping unit on a well or work on a pumping unit, you need to read and understand the following:

1. Installing a beam pumping unit or changing the parts on a pumping unit requires adequate training for individuals to perform the task safely.
2. Pumping unit installation requires heavy-duty and dependable equipment to safely lift and install the beam components.
3. All the employees are required to wear personal protective equipment (PPE):

   - Hard hats
   - Protective footwear with hard steel toes
   - Safety glasses with side shields
   - Personal H2S monitors
   - Full face shields
   - Goggles and clean clothing/coveralls
   - Fall protection
   - Rubber insulated gloves
   - Proper tools and equipment

Three Simple Methods to Install a Beam Pumping Unit

First Option

Inspect and clear the area around the wellhead. Check the vertical polished rod,

if any. A designated area at the well site will be leveled off, and a fine-grain rock base material will be placed in the form of rectangles larger than the pumping unit base beam and the cement slab of the pumping unit. Ensure that the material for the pad is flat and level. A ready-made (prefabricated) concrete cement slab, longer and wider than the pumping unit base beam, will be spaced and set on the base material near the center of the wellbore. Set the cement base and ensure that it is centered at the center of the wellhead and leveled from side to side. All the personnel are prohibited from walking or standing under any suspended load at any time.

Run a string line from the center of the wellhead through the center of the concrete base and mark the position of the main frame centers and setback. The pumping unit parts and pieces will be constructed and put together off the side, and once the unit parts are made together, the whole unit will be picked up carefully, spaced, and put in place on the ready cement slab (a heavy-duty crane may be used). Do not place any body parts between the counterweights and the crank arms at any time. (Exercise caution to prevent serious accidents.)

The pump jack base beam will be respaced and securely bolted down with special fabricated tie-down bolts and nuts to the cement slab. (The beam must be spaced out so that the carrier bar on the horse head lands over and at the center of the wellbore). Electricians will install a short wooden pole just near the back of the cement slab to install the electric control panel and the electrical meters' parts. The electric transformers will also be installed nearby, not more than 150 feet from the control panel. The electric power cable must be buried four feet underground and up to the electric panel and then to the prime mover (electric motor). The electric power cable must be buried at least four feet deep through an electrical conduit (PVC) to protect people and public transportation equipment working on location. Call and check before digging; it is the law. Check all the components for loose connections to avoid accidents.

Second Option

To avoid lifting the entire pump jack, the crew may choose to set a ready-made cement base first, followed by the pumping unit components. The base beam and gear box will be put on the cement foundation first,

followed by the Samson post and the rest of the pumping unit components. All the components of the pumping unit (pump jack) will be put together and built one piece a time.

Check the gear reducer (gear box) for the appropriate oil level before spotting. If a flush is required before filling with oil, do not use diesel, kerosene, or gasoline to wash the inside of the gear box (the gears are sensitive to bad solutions that cause corrosion). Always fill up the gear box with the recommended oil by the manufacturer. Pick up and install the gear box and level it off. Use proper torque on the tie-downs.

Install the crank arms and install the wrist pins onto the crank arms. Install the drain plugs and tighten them, and then apply a sealant. A complete unit will be made up on the special ready cement base (or a hard cement-based material). Make sure the unit is spaced correctly and the horse head is hung directly over and at the center of the wellbore.

Third Option

In the early days, the operator would mix and pour the cement base before moving a pumping unit and place it on a permanent pumping unit foundation.

Before placing the pumping unit parts, ensure that the material for the pad is level. If the ground setting is expected, it may be necessary to taper the pad slightly from the front to the back.

Make the concrete frame, pour the concrete in place, and level off as needed.. Cover up the concrete frame if necessary from rain. You must wait a minimum of forty-eight hours or longer before setting any part of the unit base. Set and align the unit as necessary. Do not use any hold-down bolts on the unit until the concrete is completely cured and hard. All the components of the pumping unit will be set and aligned as necessary.

The horse head, with the wire rope bridle and the carrier bar, must be installed properly. A throat bolt, as a safety measure, will be added to keep the horse head from shifting sideways or falling (the unit must be spaced at the center of the well). The horse head bridle must be positioned straight to hold the carrier bar flat and level at the center of the horse head. The unit will be put on pumping mode to check all the moving components. After operating on pumping mode for one month, check the pumping unit parts again for loose parts.

Important Notes regarding Pumping Unit Maintenance

The pumping unit is a unique and expensive piece of equipment. Like any other piece of machinery, the pumping unit must be lubricated and maintained properly to operate the unit safely. The pumping unit must be balanced and maintained properly to avoid serious surface and subsurface damages and to extend the pumping unit's life. Before working on a well with a beam pumping unit, you must use a "lock out and tag out" (LOTO) system. LOTO must be applied to the power source of the prime mover (gas or electrical). If the prime mover is electric, the master switch must be locked out.

Never work on or near the crank arms or counterbalanced weights until all the stored energy is isolated. Serious injury and death could occur if the crank arms and the counterbalance weights are in motion while personnel are working near the unit during workover operations. Do not depend on the brake system on the pumping unit as a safety stop. The brake is intended for operational stops only (do not misuse the brake system).

Balancing a pumping unit may require one to remove, re-spot, and/or add counterbalanced weights. An unbalanced unit is either weight heavy and/or rod heavy (the electric power surge on the up stroke and the down stroke at the electric panel will clearly indicate whether the unit is balanced and/or off-balance). The pumping unit cranks are very heavy and are often used as counterbalance weights alone if needed. The change of stroke length on the cranks may change the counterbalance. Good counterbalancing of the pumping unit is necessary to maintain minimum power surge on the up stroke and the down stroke to avoid burning the electric motor, burn the belts, and cause damage to other pieces of the surface equipment.

The rods must be checked to avoid bumping or jarring on the down stroke or hitting or tapping on the up stroke. The tail bearing, saddle bearing, and wrist pin bearings must be greased to operate smoothly and quietly without any squeaking sounds from one mile away. The wrist pins must be greased and kept cool at all times. Stop the unit when the wrist bearings feel hot. A hot wrist bearing will indicate that the bearing is beginning to fail. If a wrist bearing becomes hot while in operation, it is a sign that serious failure may take place. The bearing may become locked up, causing the pumping unit arms or walking beam to twist off (stop the unit, inspect, and lubricate). Frozen wrist bearings have caused the bearing shaft to rotate long enough to enlarge the crank bearing hole. The wrist pin bearings, tail bearing, and saddle bearing must be serviced by pumping grease up to the bearings as recommended by the pump jack manufacturer.

Metal shavings coming out of the tail bearing or saddle bearing mixed with old grease and dirt are signs of bearing failure. The squeaking sounds of bearings heard a quarter mile away must be taken seriously (you are not doing your job, friend). A twisted walking beam with broken crank arms and a bent polished rod is due to the lack of maintenance (supervision has seen before that the unit looks like a horse falling into a gopher's hole with its legs up, believe it or not).

I recommend inspecting and maintaining the oil level in the pumping unit's gear box once a year to prevent gear box damage. Filter the gear box oil at least every two years to avoid gear box breakdown. Check for sludge, water, and abrasive brass and steel shavings in the gear box oil. (Keeping up with the oil level in the gear box is the part of the operator's job.)

The Brake System on Pumping Units

Do not depend on the brakes on the pumping unit as repeating safety stops. The brakes on the pumping units are intended for operational stops only (do not misuse the brake system). The pumping unit brakes must function properly at all times to avoid accidents. The brakes must work properly to safely respace the rods and/or work on a well. The brake system on some pumping units is ignored in the oil field (unsafe workplace to operate in). The pumping unit brakes may appear differently on some pumping units. Some are bolted to the gear box.

To Start a Pumping Unit

Always clear the area around the pumping unit. Reach the brake handle

at the tail of the pumping unit. Press and release the brake slowly to bring the cranks in down position. Keep away and stay clear from swinging cranks (do not stand under the horse head of the pumping unit). When the brakes are released, the released force holding the crank arms will cause the cranks, I-beam, and head to swing with force (you never know what is going to fly at you).

Working on a well with a beam pumping unit requires a dependable brake system. Check and repair the brake system on the pumping units before moving a workover rig to a well (a good-standing brake system will save you time). Never use chains to hold the cranks upright.

Working on large pumping units without brakes is difficult and risky. The brake assembly is connected to the brake lever via a push/pull type brake cable. The cable is normally rated for a maximum of one thousand pounds. The cable should be inspected and lubricated with grease at least once a year or as appropriate.

The belt guard over the pulley sheaves and the guard around the pumping unit must be maintained and kept secure and stable at all times to protect the public and animals. The belt guard will keep the sheaves and belts dry and avoid slippage. Never get inside the pumping unit guard when a unit is in motion. The entire pumping unit must be fenced if possible (safety issue).

The Advantages of Beam Pumping

- Pump jacks are the best choice for artificial lifting in a well with low production and in the remote oil field area.
- Pump jacks are used for low-producing oil wells onshore and may be used on wells of shallow water lakes, bays, creeks, and river platforms (e.g., the Baytown and Goose Creek bays near the ship channel of Houston, Texas, and the Louisiana shallow waters).
- All the pumping units on shallow water platforms must be equipped with electrical shutdown pollution-pot devices to avoid any amount of oil and salt water leaks as well as pollution into navigable waters (one gallon of oil spill over water must be reported with fines).

Color of Spill	Gallons/Square Foot
Barley sheen	0.0000008968
Silvery sheen	0.0000017935
Bright bands	0.000007174
Dull brown	0.000021522
Dark brown	0.0000466311

$$\text{Length} \times \text{width} \times \text{gal./sq. ft.} = \text{volume of spill in gallons}$$

Beam pumping operations over the water can be pollution free. Pumping units can operate on available electricity, electric power generators, and/or IC gas engines (using propane or dry natural gas).

- When using pumping units, one may change the pump strokes to speed up or slow down the pumping unit. Changing the pump stroke length and downhole configuration is a fairly easy and safe task if it is done correctly. Changing the pump stroke length should be carried out by a professional crew with appropriate tools and equipment
- Pumping units may operate with IC engines using propane gas and/or produced wellbore dry gas (using Ajax engines).

- Pumping units are easy to install and operate as well as easy to troubleshoot and diagnose subsurface equipment performance.
- Pumping units will have lower operating costs than other artificial lifting options.
- A portable trailer-mounted pumping unit may be used to test and operate a well.
- Pumping units are the preferred artificial method for low-volume production wells.

- The equipment and service are readily available throughout the oil industry.
- The pumping unit has a longer operating lifetime (some units may last seventy years or longer with proper maintenance care).

The Disadvantages of Beam Pumping

- Too many moving parts on the surface and the subsurface
- Requires constant supervision and maintenance
- Rod wear, parted rods, holes in tubing, and bad fluid pumps expected
- Corrosion, erosion, and fatigue on subsurface equipment
- Tubing failure caused by corrosion, erosion, rubbing, sliding, and rod splits
- **Gas locking, water hammering, and fluid ponding expected**
- Surface leaks, pollution out of stuffing box, and pack-off elements
- Safety and pollution issues such as H2S and CO2 at surface leaks
- Needs fencing for safety to protect public life and property

- Counterbalancing (cannot operate the unit without proper counterbalancing)
- Tapered rod design and calculations
- Dynamometer card evaluations (waste of time)

Change of Strokes per Minute on a Pumping Unit

Changing the speed on the pumping unit (strokes per minute) can be accomplished by changing the size of the sheave on the prime mover (electric motor). Changing the prime mover sheave is fairly easy and less costly than changing the large sheave (flywheel) on the gear box. The sheave can be called a "pulley." Pulleys on the pumping units may have several belt grooves. You may select a single power band belt and/or several individual rubber belts to drive the pulleys (depends on the pumping unit size). The production volume on a beam pumping well is based on the strokes per minute, the stroke length, the downhole pump size, and reservoir deliverability.

To prolong the life of downhole and surface pumping equipment, I recommend operating the pumping units on slower strokes and with long strokes per minute. Big pump jacks should have five to seven strokes per minute only. A small pumping unit operates at eight to ten maximum strokes per minute. Without changing the speed on the prime mover, you may be able to increase or decrease the strokes on the pumping unit by changing the small pulley on the prime mover (change the driving sheaves, not the driven sheaves).

Below are some examples of pump unit speed change calculations:

Example 1

d = diameter of the prime mover sheave (pulley)	d = 6"
RPM = prime mover rotation	RPM = 1,120
D = diameter of the gear reducer sheave (large sheave)	D = 27"
R = gear reducer ratio (stamped on the gear box)	R = 36

$$\text{Strokes per minute} = \frac{\text{RPM} \times d}{R \times D} = \frac{1,120 \times 6"}{36" \times 27"} = \frac{6,720}{972}$$

$$\text{Strokes per minute} = \frac{1,120 \text{ RPM} \times 6"}{36" \times 27"} = \frac{6,720}{972}$$

Strokes per minute = **6.9**

To slow down the pumping unit speed, you need to change the prime mover sheave to a smaller diameter. To increase the speed (RPM) on the unit, you need to change the prime mover sheave to a larger size.

Example 2

d = 7" (prime mover sheave OD)
RPM = 1,175 (prime mover rotation)
D = 26" (gear reducer sheave/flywheel)
R = 28.45 (gear ratio found on the gear box)

$$\text{Strokes per minute} = \frac{1{,}175 \times 7}{28.45" \times 26"} = 11.1 \; (it \; is \; fast)$$

If you plan to slow down the pumping unit, you may have to change the prime mover sheave to a smaller diameter (from 7" to 6"). Do not mess with the large flywheel.

You actually reduce the pumping unit speed by 1.6 strokes per minute or 96 strokes per hour (less wear and tear on the equipment).

$$\text{Prime mover sheave size} = \frac{\frac{stroke}{minute} \times \text{gear ratio} \times \text{diameter of large sheave}}{\text{RPM of prime mover}}$$

If you stand in front of a pumping unit running at fifteen strokes per minute, you will have mixed feelings of running the unit fast again (some people do not know this at all). Fifteen strokes per minute at the surface will not allow the rods to make a full cycle. I prefer to operate a pumping unit on long and slow strokes per minute.

If you running a pumping unit at more than ten strokes per minute, you are running the horse to death, and it will fall to the ground soon. A large unit should be run at a slow RPM of six to seven. What is the reason that you are operating a pumping unit too fast? If you are in hurry to get a large volume of fluid out of the well to maximize your production, you may need to look for a lifting alternative.

There are several choices to increase the fluid on beam pumping:

a) Change to the longest stroke
b) Change your downhole pump to a larger one
c) Change the prime mover pulley (sheave) to speed it up
d) Change to a larger pumping unit
e) Change to another artificial lifting system

You must use safety guards around the pumping unit to protect people and animals (you do not want to buy an expensive dead show cow). Never operate a pumping unit without safety guards (keep safe). You may often see a pumping unit down and not going up and down.

The main reasons for the pumping unit's downtime are the following:

1. The well is working on a time clock (most often, low-producing or stripper wells are put on time clocks, waiting on the reservoir fluid to rise above the pump to avoid fluid pounding)
2. Electrical problems (burned fuses, power supply, or transformer)

3. Surface equipment problems (prime mover, drive belts)
4. Holes in the tubing string (caused by corrosion, rubbing, sliding, wear)
5. Parted sucker rods (fatigue, bad makeup, bad design, wear)
6. Subsurface fluid pump not lifting fluid (caused by sand, trash, gas locking, and leaks)
7. Sanded-up rods and perforations (formation sand, mud from perforations)
8. Broken polished rod (pounding, tight clamp, fatigue, bad makeup)
9. Stuffing box leaking (worn packing, hot polished rod, poor lubrication)
10. Pumping unit problems (tail bearing, saddle bearing, gear box)
11. Well flowing (new plug completions)
12. Temporarily abandoned wellbore (any serious downhole problems)
13. The price of oil is too low (it just does not pay off)

CHAPTER

SUBSURFACE FLUID PUMPS

How to Calculate the Subsurface Pump Displacement

The production calculations are based on the theoretical capacity and the actual pump capacity. Pump bore multiplied by the length of plunger travel will give you cubic inches of fluid per stroke (**231 cubic inches = 1 gallon**).

Example I

Pump plunger 2.25″ = 2 ¼″

Strokes per minute = 8

Stroke length = 160″

Pump plunger constants:

Size	Factor
1 ¼″	.182
1 ½″	.262
1 ¾″	.357
*** 2″	.468
2 ¼″	.59
2 ½″	.728
2 ¾″	.88
3 ¼″	1.231
3 ¾″	1.64

Theoretical Pump Displacement Calculation

24 hours of fluid displacement = strokes per minute × stroke length × pump constant

Example I

Pump stroke = 8 strokes/minute

Stroke length = 160″

Plunger size = 2"

24 hours of pump displacement = 8 × 160" × 0.468 = 599 barrels (bbl)/day at 100% efficiency

755 × 80% = 479 bbl/day (may be closer to actual daily production)

Real production is the actual fluid displacement in twenty-four hours when you gauge the oil stock tank. The pump calculation consists of the number of strokes per minute, the stroke length, and the size of the pump plunger (less stretches and over/under travel).

Example II

Strokes per minute = 8

Stroke length = 160"

24 hours of pump displacement = (plunger diameter)2 × SPM × stroke length × .1166

24 hours of pump displacement = 2" × 2" × 8 × 160 × .1169 = $599\frac{bbl}{day}$ x $(100\%\ efficiency)$

$599 × 0.80 = 479\ \frac{bbl}{day}$ x $(80\%\ efficiency)$

Volumetric pump efficiency = $\frac{actual\ volume}{theoretical\ volume}$ x 100

KHOSROW M. HADIPOUR

The Major Components of Subsurface Artificial Fluid Lifting

Do you know how a subsurface fluid pump works? Pay attention to the following pages carefully. Before talking about subsurface fluid pumps, we need to describe the important tools and equipment that make the fluid pump operation effective and efficient.

Major Subsurface Fluid Lifting Equipment

1) Tubing String

The tubing string is one of the major necessities and the most important part of any type of artificial lifting system in oil and gas operations. The size and grade of the tubing that is implemented in artificial lifting is based on the well depth and wellbore mechanical integrity (MIT) (tubing sizes range from 2 ⅜" OD to 4 ½" OD). Tubing strings are normally used in flowing or artificial lifting applications.

2) Tubing Anchor Catcher (TAC)

The TAC is used to anchor the tubing string and catch parted tubing.

3) Conventional Subsurface Fluid Pumps in the Oil Field

   ✓ Insert pumps (various types and sizes)
   ✓ Tubing pump (known as the working barrel)

Seating mechanism :

   • Mechanical seating nipple
   • Standard API seating nipple

4) Sucker Rods String

- Steel sucker rods (various size and grades)
- Fiber rods (various sizes and grades) [insert 00169]

All of the above subjects will be explained in the following pages.

Section I

Oilfield Production Tubing String

Regarding the downhole tubing string (production string) in oil and gas operations, the pipe is also called the tubing. Actually, the oil field pipe is defined as a closed cylindrical hollow conductor that is made from steel or high-pressure fiberglass and is used in oil and gas wells to transport oil, gas, or liquid from one point to another.

Tubing may come in different shapes other than round pipes. Tubing may be round, square, and/or rectangular. In the oil field, we may call tubing a pipe and/or may call the pipe a tubing joint. We have not used square pipes in oil wells as yet (it simply will not work). The difference between the pipe and tubing may also be the wall thickness.

The pipe or tubing is manufactured and come from all over the world. Some foreign manufactured pipes are no good (need lots of improvement). The oil well casing and tubing string is the most valuable component of all the artificial lifts or flowing wells in oil and gas operations around the world.

The steel casing and tubing string is the major source of consumption in oil and gas well operations. The success and failure of all oil and gas wells depend on good, reliable tubing and casing strings. Do not apply inferior-quality pipes to oil fields. The economic impact of good or bad manufactured tubing or casing is enormously high in oil and gas operations. API tubing and casing is recommended for all oil and gas wells. The manufacturing of steel products such as casing and tubing is unique, with highly technical and scientific engineering and skill requirements.

How Pipes Are Made

The production process of casing and tubing is derived from a raw iron material that comes from iron ore, coke, limestone, and other materials that are burned at high temperature in a furnace to produce "pig iron." Pig iron is the main raw material in steel production and contains several elements such as carbon, silicon, manganese, phosphorous, and sulfur. Pig iron will pass through different heating and refinement processes to make various steel pipe products to meet oil and gas operation requirements.

Steel tubing is subject to unbelievably harsh and hostile wellbore environments. The required mechanical properties of steel casing and tubing include the following:

- Tensile strength — The maximum stress load that tubing can hold before rupture.
- Yield strength — The value of maximum stress to strain. The yield strength is the applied stress that describes the load above by which a material may be deformed permanently.
- Hardness — The mechanical resistance of the material to deformation. Hardness is a mechanical property of metal to resist indentation or penetration.
- Ductility — The mechanical resistance to rupture. Ductility is the measure of the body to elongate or stretch before rupture (deform plastically without fracturing).
- Performance Properties of Tubing and Casing for Oil and Gas Wells
- Collapse — The resistance of tubing to withstand external pressure; applied force to the outside of the tubing that causes the pipe to cave in.
- Burst — The ability of tubing to withstand internal pressure.
- Tension — Related to the axial stress, which is the ability to withstand torque, tension, and compression (tension is the maximum pull before the pipe becomes parted).

Chemical Elements Present in Steel Pipe Making

- Carbone — The carbon content in the steel product is used to classify the types of steel products. Carbon increases hardness, tensile strength, and abrasion resistance. Too much carbon will reduce ductility and toughness. Higher carbon content is used for cutting tools, slips, dies, and files (the relation between carbon and steel can be described like that between salt and water).
- Manganese — Increases hardness, abrasion resistance, and tensile strength and decreases ductility.
- Silicon — Increases hardness, tensile strength, and deoxidization.
- Aluminum— Increases deoxidization and strain aging.
- Chromium — Increases hardness and corrosion resistance.
- Phosphorous — Increases tensile strength and hardness, decreases ductility, and improves machinability.
- Sulfur — Increases machinability and hardness and decreases weldability, ductility, toughness, and corrosion resistance.

Other elements may be added to improve the physical and chemical characteristics of steel tubing. Each of the above additives will have a positive effect on the physical and chemical conditions, tensile strength, hardness, ductility, and yield strength of the manufactured pipes. Starting from raw ore material to a finished tubular product will require many engineering processes.

There are two types of manufactured tubing/casing in oil patch operations:

A. Seamless tubing (SMLS) (without seams)
B. Seam-type tubing (ERW)

A-The Manufacturing Process of the SMLS Pipe

The SMLS pipe is made from a solid steel cylinder. This is referred to as a billet, which is heated and pierced into the pipe without any welding or seams. The SMLS pipe is generally stronger and more expensive than the ERW pipe.

The round steel billet is solid semi-finished round or square steel that is cast from a forging process. The billet will pass through a heater/furnace of high temperature until it is near melting. From the furnace process to the elongator piercer, the hot tube will pass through several rollers to achieve a satisfactory wall thickness and smooth physical shape. The tube will continue passing from the elongator piercer to the furnace and to the hot stretch reduction mill to form the outside diameter (OD). The tube will then be cut into pieces and cooled off. The austenite heating process is a process that evaluates the tube temperature above 1,500 degrees to "austenite" until all the elements are in the solution before the cooling process and quenching.

From the Heat Treat Process to Quenching

Quenching and tempering are used to alter the microstructure of the tubing to improve its strength. Quenching and tempering will control hardness, reduce brittleness, and bring the steel to its required tensile and yield strengths (J-55, P-110, N80, L80). The pipe will then goes to straightener machines and undergo testing. After testing, the tubing will pass through the length cut and undergo threading. The coupling will be put on, and the pipe will be drifted. The tubing will be hydrostatically tested depending on the grade. Final testing will be conducted, and pipe protectors will be installed.

Important Steps in Tubing Manufacturing

- Heat treatment
- Quenching
- Full-body normalizing after upset (a very important step)
- Full-body induction
- Stress relieving
- Ductility test
- Elongation test
- Yield strength test
- Tensile strength test
- Hardness test

Seam-Type Welded Tubing Joints (ERW)

Basic manufacturing processes are used for making near-perfect seam-type tubing from a flat coiled steel sheet. Tubing manufactures produce welded steel pipes by forming a flat steel sheet into cylindrical structures and closing the joint edges via welding:

a) Furnace weld (FW) or continuous weld (CW)
b) Electric resistance weld (ERW)

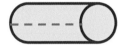

Most oil field casing and coiled tubing are manufactured as ERW pipes (the ERW seam is visible in the smaller tubing). Careful consideration is needed when testing and running the ERW casing in a well to avoid splits and ruptures. (Avoid major complicated casing problems in the oil strings before it is cemented in the borehole.)

Typical Manufactured Oil and Gas Tubing and Casing Joints:

Tubing Data

Tubing OD	Weight	Inside Diameter	Capacity (bbl/ft)
1"	1.80#	.073"	.00107
1 ¼"	2.40#	.970"	.00185
1 ½"	2.90#	1.210"	.00252
2 ⅜"	4.70#	1.995"	.00387
2 ⅞"	6.50#	2.441"	.00579
3 ½"	9.2#	2.992"	.00871
3 ½"	9.30#	2.992"	.00870
4"	9.50#	3.423"	.01223
4"	11.0#	3.476"	.01175

Casing Data

Casing OD	Weight	Inside Diameter	Capacity (bbl/ft)
4 ½"	9.50#	4.090"	.0162
4 ½"	10.50#	4.052"	.0159
4 ½"	11.60#	4.000"	.0155
5"	15.0#	4.408"	.0188
5"	18.0#	4.275"	.0177

5″	21.0#	4.154″	.0167
5 ½″	13.0#	5.044″	.0248
5 ½″	15.50#	4.950″	.0238
5 ½″	17.0#	4.892″	.0232
5 ½″	20.0#	4.778″	.0222
7″	17.0#	6.538″	.0415
7″	20.0#	6.456″	.0405
7″	23.0#	6.366″	.0393
7″	26.0#	6.276″	.0384
7″	32.0#	6.094″	.0362
7 ⅝″	20.0#	7.125″	.0493
7 ⅝″	24.0#	7.025″	.0479
7 ⅝″	26.40#	6.969″	.0472
7 ⅝″	33.70#	6.765″	.0444
9 ⅝″	36.0#	8.921″	.0774
9 ⅝″	40.0#	8.835″	.0758
9 ⅝″	43.50#	8.765″	.0765

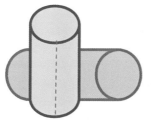 

Oil Field Manufactured Tubing Descriptions

➤ EUE Tubing — Referred to as the "external upset end."

For example: 2 ⅜″, EUE, 8 round, J-55, R-2, T&C tubing

This means: 2.375″, external upset end, 8 round pipe, J-55, range 2, tubing thread and collar (8rd means 8 threads per inch)

➤ NUE Tubing — Referred to as "non-upset" tubing joints (non-upset end)

For example: 2 ⅜″, NUE, 8 round, J-55, R-2, T&C tubing

This means: 2.375", non-upset end, 8 round threads, J-55, range 2 pipe, tubing and collar

➤ Integral Tubing Joints — Of various sizes, integral joints are manufactured with male pins and female boxes, which are the most integral parts of the tubing (a special MYT or YT elevator must be used to run this pipe).

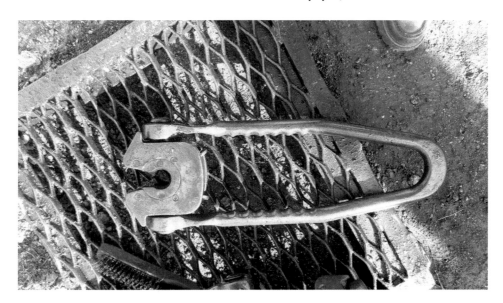

➤ NUE Tubing — PH-6 P110 tubing

may require special elevators and the handling of tools such as YT or MYT to pick up and/or lay down tubing.

➤ Tubing Collar — Made up from an SMLS pipe and threaded internally from the top to the bottom. Collars may be referred to as tool joints, where the two joints of a pipe are connected. Consult with the tubing manufacturer before high tension on tubing occurs.

Tubing Size (Seamless)	New Seamless Tubing (Safe Pulling)
2 ⅜" EUE, 8rd, 4.70# H40 tubing (new)	50,000 lbs. max
2 ⅜" EUE, 8rd, 4.70# J-55 tubing (new)	70,000 lbs. max
2 ⅜" EUE, 8rd, 4.70# N80 tubing (new)	90,000 lbs. max
2 ⅞" EUE, 8rd, 6.50# H40 tubing (new)	70,000 lbs. max
2 ⅞" EUE, 8rd, 6.50# J-55 tubing (new)	90,000 lbs. max
2 ⅞" EUE, 8rd, 6.50# N80 tubing (new)	120,000 lbs. max
3 ½" EUE, 8rd, 9.30# H40 tubing (new)	90,000 lbs. max
3 ½" EUE, 8rd, 9.30# J-55 tubing (new)	130,000 lbs. max
3 ½" EUE, 8rd, 9.30# N80 tubing (new)	180,000 lbs. max

The American Petroleum Institute (API) specifies a variety of tubing and casing threads applicable in the oil field. Casing threads can be short threads (STs), long threads (LTs), or buttresses for casing strings. These threads may be called interference connections given the fact that the connection is sealed by the wedging action of two tapered surfaces coming in contact. A modified thread lubricant is used in the voids for a better sealing element.

The eight round (8rd) threads are the most popular in tubing strings. The term "8rd" is derived from eight threads per inch with a rounded cone-shaped figure. Buttress threads and 8rd threads are tough threads if used properly. Buttress threads are nearly square in shape. The engineering appearance will make the buttress threads much stronger than the 8rd threads when under tension. The thread tolerance on buttress threads is much higher than that on 8rd threads. EUE 8rd threads have more resistance against fatigue. Also, 8rd tubing is the most popular type of tubing used in the oil field because of its higher quality and tripping speed.

End of Thread Beveling

There are two types of tubing thread ends: square end cuts and beveled end cuts. I prefer the bullet-nosed "beveled thread end" than the square end on production tubing joints.

Typical Tubing Used in the Oil Field

- Carbon steel tubing
- Stainless steel tubing
- Carbon moly tubing
- Chrome moly tubing
- Copper nickel tubing
- Monel tubing
- Nickel tubing

Tubing quality is very important; it can save you costs, especially in offshore operations.

Major Causes of Tubing Failure in Oil and Gas Operations

1. Mishandling tubing by people (a major problem from the factory to the wells) (e.g., dropping, beating, dragging, tossing, twisting, tubing makeup)
2. Thread galling (Over-torque makeup without an adequate thread compound will result in thread galling. Galling can occur when using a backup device on the tubing collar in making up the joints. Tubing backups on the tubing collars may cause galling (wrong spot). Repeated joint makeup during fishing can cause interference problems.)
3. Corrosion (H2S and CO2 corrosion is the number-one enemy of tubing)
4. Temperature and abnormal pressure
5. Wrong application of tubing during stuck pipes, fishing, and jarring (use work string)
6. Rod wear (rubbing, sliding, rolling, and erosion in crooked holes)
7. Bad tubing makeup (improper makeup of under-torque or over-torque tubing)
8. Poor lubrication before makeup
9. Misuse and wrong application of tubing string (application of the pipe in fishing, corrosion treatment, chemical stimulations)
10. Dropping the tubing in the well without any fluid
11. Improper lubrication on the pin and collars during the tubing makeup
12. Manufacturing defects (heat treatment, quenching, normalizing)
13. Wrong tubing landing depth (landing tubing string across and/or below open perforations; you will increase external holes on the tubing string)
14. Tubing damage during high-pressure stimulation
15. Using the pipe on snubbing and stripping
16. Application of tubing string in corrosive environments without proper treatment

Corrosion is the loss of identity in oil field equipment without chemical treatments in H2S wells and CO2 wells. Corrosion may result from an electric current or galvanic reaction between dissimilar metallic members (such as steel alloy and carbon steel). With proper care and proper chemical treatments, most tubing failures can be prevented.

Care and Handling of New Tubing String

The number-one enemies of tubing and downhole equipment are (a) misuse and misapplication (running and pulling the tubing string) and (b) a downhole corrosive environment without good chemical treatments.

- Unload tubing string with care. Use forks to load and unload tubing joints.
- Do not drop or slap the pipes together when unloading or lifting them.
- Avoid tubing bend. Discard any and all bent tubing (damaged pipes).
- Avoid using tubing for the wrong applications to save money (fishing, drilling).
- Repeated trips such as fishing and jarring will cause the tubing string to deform and break at the last makeup thread on J-55 and N80 tubing. Do not use J-55 or N80 tubing on extensive fishing operations.
- Drift tubing when ready to lift and run the pipe. Discard tight drift pipe.
- Use aluminum or fiberglass rabbits in coated tubing strings only.
- Keep the rig block and elevator at the center of the wellbore to avoid tubing damage. Avoid running tubing during storms and high winds.
- Do not remove the thread protectors from the tubing ends until you are ready to stab the pipe on the rig floor.
- Keep the pin and collar threads very clean and lubricated before makeup.
- Remove the tubing protectors just before making up the pipe.
- Avoid using a dirty lubricator with dirt and sand. Use a special soft brush.
- Apply a light, fresh, and clean thread compound over the entire pin threads using a brush utensil just before stabbing (use a lead-free thread compound).
- Make up the tubing in three complete rounds using a hand or pipe wrench first. Then use power tongs to make up the tubing (this will prevent thread galling).
- Check the makeup torque of the pipe every five hundred feet to avoid under torque/ over torque. Apply the recommended makeup torque as appropriate.
- Do not mix stainless tubing with carbon steel tubing (there is a 90% chance this will cause heat and galling problems when backing off).
- The tubing should be torqued up as recommended by the tubing manufacturer.
- Check and adjust the hydraulic torque pressure on the power tongs every fifty joints.
- Check the slips and tong dies to avoid damaging the tubing body.
- Do not forcefully drop the pin into the tubing collar before making up the pipe; tripping it will cause a hammering effect and damage to the threads.
- Never use hammer blows on the coupling to break a tool joint.

- Complete back-off turns are necessary before pulling the pin out of a collar.
- Hammer blows will permanently damage the threads (causing thread leaks).
- Use protectors when pulling and standing back tubing string (prevents sand and solids from entering pin threads).
- Use protectors on all the PH-6 working strings and the buttress pipe when tripping and standing the tubing string out of the well.
- Use a stabbing guide for all PH-6 tubing, casing, and buttress tubing joints.
- Threads on the pin end and collars must be kept very clean when pulling and running the tubing string.
- All the tubing joints with external and internal corrosion pits and rod wear must be removed from the string to avoid losing the tubing in the well.
- Apply a clean thread compound on the pin end only (the light pipe dope must cover the entire thread area).
- The thread compound must be kept covered and free of sand at all times.
- Note that the foot pound thread makeup torque valve is different from the psi makeup torque value (approximately 1 psi = 1.365 ft./lb.).
- Pay attention to worn threads on the tubing pins and couplings to avoid dropping the pipe into the hole and cause extensive fishing costs.
- Keep the pipe elevator in good and safe working conditions to avoid dropping the tubing into the hole and causing possible accidents.
- If the tubing pin threads

drop deeper into the tubing collar

during pipe makeup, there is something wrong with either the pin or the collar threads. Stop, clean up, and check the pin threads and the collar before making up the pipe. Galled threads will cause the pipe to leak and become parted in the well.

The pipe stretch may be calculated using the formula below:

$$L = \frac{S \times 1{,}000 \times 1{,}000}{P \times C}$$

L = estimated length of free pipe in the well above the stuck point
S = amount of stretched length pulled above neutral weight
P = total tension pulled into the tubing string
C = constant stretch valve of the stuck tubing per 1,000′

Example:

2 ⅞″ stuck above a packer at 8,000′

    a) Zero the weight indicator on the rig.
    b) Get the tubing to neutral weight and mark the pipe at the slips.
    c) Pull 20,000 lbs. above the weight of the tubing string (say 65,000 lbs.).
    d) Mark the pipe again at the slips.
    e) Measure the distance between the two marks (say 30″).

f) L = $\dfrac{30'' \times 1{,}000 \times 1{,}000 = 30{,}000}{20{,}000 \times .22 \quad\quad = 4.4}$

g) L = 6,818' +/− (pipe may be stuck at or near 6,818')

Note: Free-pointing stuck casing and tubing is complicated and may require professional electrical wireline tools and equipment for an accurate free point or stretch reading survey before cutting or backing off the tubing string. I will describe the free point, back-off, jet cut,

and chemical cuts later.

Testing and Inspecting Methods of Tubing and Casing in the Oil Field

- Visual inspection — The pipe must be clean and free of dust and dirt with good vision.
- Mechanical gauging — Drifting and measuring outside and inside of the pipe.
- Electromagnetic testing — Scanning (limited accuracy and a not-too-reliable method).
- Eddy current testing — Electromagnetic testing to detect cracks.
- Ultrasonic and/or gamma ray testing — Using radioactive isotopes.
- Hydrostatic testing — May be conducted using water pressure and/or gas testing (dependable method with limitations).

The above testing methods are to find leaks, corrosion pits, cracks, holes, and surface imperfections. Read on hydrostatic and hydro-testing in oil and gas by this author.

Electromagnetic Testing and Coding Method in Oil Patch

The applications of tubing color-coding identifications are as follows:

- 0% to 15% wall loss (85% of pipe body remaining) — Yellow band code (yellow band pipe is good)
- 16% to 30% wall loss (70% of body wall remaining) — Blue band code (Blue band is good for flow line)
- 31% to 50% wall loss (50% of body wall remaining) — Green band code (Green band tubing is good for structural use only)
- 50% to higher wall loss (less than 50% wall remaining) — Red band code (Red band pipe may be used as structural pipe)

Oil field tubulars must be check and drifted to obtain a full inside drift diameter. Tubing and casing must fully be drifted prior to running in a well (discard all the casing and tubing with tight drift diameters and oval/egg-shaped bodies). Never misuse the tubing string. The misapplication of tubing in drilling, milling, snubbing, stripping, and fishing operations will cause the tubing to fail (do not use the wrong tubing string for fishing and snubbing jobs). The thicker the pipe, the smaller the inside diameter (ID). In the oil field, we purchase pipes or tubing based on the nominal pipe size (NPS), referred to as the OD: 2″, 2 ⅜″, 2 ⅞″, 3 ½″, and 4″. The nominal size sometimes may be referred to a pipe according to its weight and grade.

Purchasing a New Tubing String for a Particular Wellbore Depth Range

H40 and J-55 tubing are designed for shallow to mid-range depths. H40 and J-55 tubing are good for corrosive fluids. For example, you may need 2 ⅞″, 6.50#, 8ʳᵈ, R-2 (H-40, J-55,

N80, L80, and/or P110 tubing). The pipe you are ordering to use in a well is either SMLS or ERW. You may use internal plastic coated (IPC) tubing.

Note that most casing strings are generally ERW

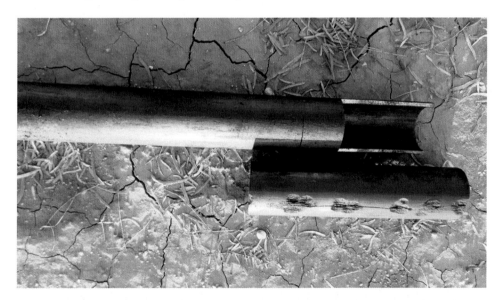

pipes (seam-type pipes). When a new tubing or casing arrives on location, it is fairly easy to identify the size and grade because of the fresh color coating and the stamp on the pipe.

- J-55 tubing is green band.
- K-55 is green with double band.
- H-40 is black band.
- N-80 is double red band.
- L-80 is red with three bands.
- P-110 is white band.

Look for a bad joint. Often few joints of different grades of pipes may be accidently mixed with other grades of pipes in the supplier's yard and/or in the fields, and it will be difficult to differentiate the right and wrong materials for the well (mixed tubing grades are subject to fail in the hole). I recommend cleaning, inspecting, and drifting all the casing strings on the well location before running the pipe into any open hole to avoid tight drifts and splits in the pipes (this will avoid possible failure/disaster later). (If a default casing is run and cemented in the hole, it will be subject to leaks and cause major failure.)

When the tubing joints become old, it will be difficult to differentiate the pipe grades. The application of mixed tubing strings becomes risky in case of emergency. (You can never tell with old pipes: J-55, K-55, N80, L80, or H-40.) The yield strength and tensile strength of each grade of pipe is different. Make certain you do not run mixed grades of tubing in any well.

Pipe weight = (OD of pipe − wall thickness) × (wall thickness × 10.69)

$$\text{Wall thickness} = \frac{\text{pipe OD} - \text{pipe ID}}{2}$$

In the oil industry, we also add to the description of the pipe length in terms of "range":

- R-1 pipe: 16' to 25' long (line pipe)
- R-2 pipe: 25' to 35' long (tubing/casing string)
- R-3 pipe: 35' to 45' long (casing string)

Used tubing joints must be stored on pipe racks according to size and grade.

Example:

2 ⅜" EUE, 8rd, J-55 T&C tubing

2 ⅜" EUE, 8rd, N80 T&C tubing

2 ⅞" PH-6, P110 tubing

Application of Varnish and Coated Material Outside and Inside of Tubing

All the tubing strings used in mechanical artificial lift should be plain and cleaned internally and externally. The tubing must be free from foreign material to prevent subsurface pump problems.

- On beam pumping wells (trash and junk will get in through perforated intake subs)
- On hydraulic jet pumps (trash may circulate down the hole and into the pump)
- On hydraulic fluid pumps (trash may circulate from the surface down into the pump)
- On electric submersible pumps (trash can get to the pump through fluid intake)
- On PCPs (trash will fall from the annulus into the rotor and stator)
- On gas lift system (loose trash and varnish may get lodged into the operating valves)

All the varnish should be removed from outside the tubing before running the pipe in a well. Varnish is used in the pipe supplier yard to protect the tubing from rust and environmental corrosion. Varnish will become soft and will break off in contact with hydrocarbon gas, condensate, and oil. It will fall off and get lodged into the subsurface pumps and chokes. It will be difficult to scrape or remove dry heavy pipe varnish from the tubing string.

Tubing with outside wrapped material

should not be used in any artificial lifting wells (what are you trying to do?). If wrapped material is used, you will cause more downhole problems. The application of IPC in the tubing may help slow down corrosion, and it may depend upon the wellbore condition and application method. Some tools may not go through coated tubing!

There are many types of coating material used to protect the pipe from corrosion and/or wear:

- Epoxy material
- Primer
- Lacquer varnish
- Rubber liner (works great in salt water or injection wells)
- Cement

Epoxy is one of the elements mostly used in the oil patch tubing string. The application of epoxy is a good practice in flowing wells, gas lift wells, submersible wells, injection wells, and/or salt water disposal wells (note that using coiled tubing and sucker rods will destroy the coating because of rubbing and sliding). The epoxy coating is either a powder or liquid coating and is an excellent idea to slow down H2S or CO2 corrosion. The applied epoxy coat should have consistency throughout the pipe and be **"holiday free"** (without air bubbles under the coating). Epoxy should be applied with care and must be holiday free.

Tubing collars are manufactured from SMLS pipes. Tubing collars are threaded through the collar from the top to the bottom internally. The tubing collar must be internally epoxy coated heavily at the void where the two member pipe joints meet (internally and at the center). Improper collar coating will create corrosion holes at the center of the threaded coupling, which may cause tool joint failure. The application of heavy epoxy in the beam pumping tubing string is a waste of money (they just do not last long enough because of rod erosion and wear in the crooked holes). A thick epoxy coat will reduce tubing drift and may keep pump plungers and/or fishing tools from passing through in case of emergency.

The application of thick varnish coating and/or the outside wrapped material is a bad practice in any artificial lifting method. These elements will cause downhole mechanical problems such as parted rods, stuck pumps, and chocking problems. All the new tubing must be protected against corrosion, erosion, and paraffin buildup. The application of dependable rod guides

and non-metallic paraffin cutters is recommended.

Tubular Failure in the Oil Field

For many years, the oil and gas industry has been searching and researching for methods to prevent steel sucker rod and tubular failure because of corrosion and wear without significant results. The application of various costly chemicals down the wellbore and into bare steel tubing strings has not been a successful approach in solving or slowing down corrosion problems. Frustrating tubing failure and the price of repeating workover costs make you wonder when some of the low productive wells will ever make revenue.

Changing the bare tubing string to a plastic coating material inside and outside has not yielded satisfactory solutions because of several factors:

- Quality and application methods
- Quality of service found to be poor

The bare tubing string costs the oil and gas industry given the lack of performance in corrosive wellbores (the tubular material is not the same quality as it used to be). Seamless

IPC tubing is cost effective and will perform well in oil well, salt water disposal well, injection well, electric submersible, gas lift, gas well, and hydraulic or jet pump artificial lifting operations as well as other applications that do not cause wear from sliding and rubbing.

Quality IPC tubing will perform well and last longer in offshore and onshore oil and gas wells with the following restrictions:

- The IPC coating must be holiday free and must pass quality control.
- The tubing must be bullet nosed to slow down fluid cut at the tip of the tubing.
- Loading and unloading the IPC tubing must be carried out with extra care to prevent damages.
- Most IPC tubing damages will take place while loading, unloading, and handling the tubing from the mill to the well.
- Forklift loading and unloading must be carried out with an experienced forklift operator to avoid denting, pushing, dragging, and dropping the tubing joints.
- No pyramid tubing loading should be allowed to prevent tubing damages.
- The pipe must be loaded and separated by divider boards.
- All the tubing joints must have tubing pin protectors and coupling protectors.
- All the tool joints must be cleaned properly at the plant, and a light clean thread compound must be applied to all the pin and collar surface threads' areas.
- The tubing must be unloaded and placed on pipe racks properly using a forklift.
- The pipe must be lifted off a catwalk to prevent slapping against the rig floor.
- Avoid dragging any joint of tubing on the ground.
- Adjust the traveling block so that the elevator lands at the center of the hole.
- The thread protectors must be removed at the rig floor only just before making up the tubing joints.
- Use a plastic cup-type stabbing guide before stabbing and making up the tubing.
- Always make up the tubing joints by hand for at least three rounds before applying hydraulic tubing tongs.
- Avoid running the tubing during high winds to avoid galling and cross-threading.
- Always apply a light and clean thread compound to the pin ends (only when necessary).
- Run full-size drift tools (use the Teflon rabbit only).
- Apply the correct and recommended manufacturer makeup torque (check the makeup torque every twenty joints caused by changes in temperature and torque value).
- Notice the difference between ft./lb. torque vs. psi-applied torque.
- Plastic-coat the bottom hole assembly (such as the packer, gas lift mandrels, and seating nipples) to prolong the useful life of IPC tubing.

There are several valuable practices to consider when running and pulling the IPC tubing string:

- Avoid running sharp objects through the IPC tubing string after installation.
- No coiled tubing should be run through the plastic coated tubing.
- No wireline broach and cutting tools should be run through the plastic coated tubing.
- Avoid dropping heavy steel bars through the plastic coated tubing.
- Do not use slip-type (drag tool) and/or hydrostatic tools through the plastic coated tubing.
- Do not use steel swab cups in the plastic coated tubing string.
- Use a Teflon rabbit only to drift plastic coated tubing.
- Do not use plastic coated tubing in cement or formation drilling.
- Do not use plastic coated tubing in fishing and jarring operations.
- Do not use plastic coated tubing to bail sand and abrasive solids.
- No snubbing or stripping is allowed.

Application of Corrosion Chemical for Downhole Treatments

There are **not** too many experts in any oil and gas business who actually understand and know enough about the chemical treatments that are applied in their oil and gas wells. Have you noticed that the sales of various chemicals are growing every year? Chemical treatment methods and chemical applications have not been changed in sixty years. Almost 70% of all the wells in the United States are sour in nature. Almost all the oil and gas wells in the world contain H_2S or CO_2 in their solutions.

The major problem in oil and gas wells is the production of corrosive water, corrosive oil, and gas with bacteria along with scale problems and abrasive formation sand production. The bare steel pipe is subject to corrosion faster than anticipated. The older version of the manufactured tubing string appears to last longer because of the integrity and quality of the steel manufacturing products. With cheaper-quality tubular products shipped in from all over the oil and gas industry, you will see daily tubing string and rod string failure after failure.

I cannot see how the oil and gas industry makes any profits out of stripper wells. When pulling the tubing string out of well, fairly new tubing becomes parted repeatedly just below the tubing upset; this is a plain joke of manufacturing tubing strings (bad heat treatments and normalizing). To slow down the rate of corrosion in downhole tubulars, some oil and gas companies use IPC tubing, as explained before, or they use "poly" insert liner tubing strings.

The plastic poly tube is exerted into steel tubing to cut IPC costs and to reduce corrosion damage internally. Steel tubing with poly will perform very well in flowing wells, injection wells, and salt water disposal wells (some may last over ten years in a well). Since 1985, poly pipes have been inserted and glued into steel tubing to slow down internal corrosion and reduce the rate of corrosion in the steel tubing string.

Poly is a product that is derived from plastic. The plastic product market is everywhere throughout the world (including a place in the Zagreus Mountains of Persia). In some sense, plastic has become a convenient object in life and been considered as an element of progress. On the other hand, plastic has become one of the major sources of pollution and hazard chemical products around the world (it is a chemical product). Nasty flying plastic bags, cups, bottles on the sides of roads that stick to fences and trees, and poly plastic tubing across oil fields like moving snakes are some of the main causes of damage to the environment that we live in.

From fancy buttons on your shirt to plastic bags, plastic sacks, plastic cups, boxes, coolers, car components, safety glasses, highway construction cones, ships, boats, bottles, jars, toys, gloves, shoes, lamps, poly tubing, computer products, phones, tires, mats—I'll let you to name the rest of the plastic products (looks like we have made progress). Plastic products are vastly used in industrial products, army and agriculture equipment, the electrical, plumbing, and oil field industries, dental application, power cables, and even artificial joints in the human body.

The major sources of plastic are hydrocarbon oil, gas, corn, sugar cane, coal, and trees. Poly plastic is made by a chemical process. Crude oil and hydrocarbon gas contribute major parts of this process: fuels, chemicals, and thick liquid polymers converted to major plastic products in the world. Plastic has a great quality and can be made in different shapes and forms, including fiber, coating, and/or glue. Plastic will have unbelievable applications in industrial, electrical, commercial, and oil field operations. Plastic derived from polymers can be called "poly." The chemical components of polymers are carbon, hydrogen, nitrogen, oxygen, phosphorous, silicon, and others. All plastic products contain chemicals. All plastic products will leach chemicals, are harmful to human health, and might cause cancers.

Poly can be manufactured as improved finished products, referred to as polyethylene tubing products. Polyethylene tubing strings are designed based on various applications and can be defined as the following:

- High-density polyethylene (HDPE)
- Low-density polyethylene (LDPE)
- Medium-density polyethylene (MDPE)

The polyethylene pipe is referred to as the HDPE pipe. It is lightweight, flexible, and noncorrosive and can be spooled for transportation easily. Polyethylene tubing is applied in oil and gas operations as production gathering flow lines to convey oil, gas, and produced water to tank batteries. The polyethylene pipe is advanced and can be used as liner tubes in oil and gas wells to slow down corrosion attacks in steel tubing strings:

- Salt water disposal wells (works very well)
- Salt water Injection wells (works very well)
- Electric submergible wells (works well)
- Artificial fluid lift of rotary pumps (do not recommend)
- Artificial lift beam pumping wells (do not recommend)
- Oil and gas flow line systems (works very well)

The polyethylene liner is basically abrasion resistant, internally smooth, and corrosion resistant and can be used in oil and gas wells with moderate H2S, CO2, and chemical applications. The polyethylene tube is made up from seamless polyethylene and mechanically bonded inside steel tubing of various sizes. The tube is lightweight, slick, and smooth, will last longer in the well, and will offer great cost saving over a bare steel tubing string (based on my experiment).

The Advantages of Poly Pipes in Oil and Gas Operations

- Reduce corrosion better than steel tubing alone
- Cost-saving measure by reducing frequent well workover caused by corrosion leaks
- Reduce tubing and equipment replacements because of holes in tubing
- May decrease pressure drop friction and wear
- Mitigates tubing and rod string wear

- Rod guides may or may not be needed if applied in sucker rod beam pumping

The poly pipe is susceptible to the following wellbore conditions:

a) Downhole temperature ratings
b) Downhole pressure applications
c) Severe downhole corrosion application in H2S and CO2 wells
d) Mechanical activities such as drilling and fishing, coiled tubing, and a high concentration of acidizing may damage poly lines and cause failure
e) Tube may wrinkle because of high temperature (unglue or block off the pipe drift)

Unlike steel tubulars, the poly pipe is not subject to corrosion. The poly pipe will resist corrosion and burst while glued inside the steel tubing string.

The following polyethylene (HDPE) liner sizes are available:

Tubing Size	Pipe Drift	Liner Weight (lb./ft.)
2 ⅜"	1.60"	0.33 lbs./ft.
2 ⅞"	2.00"	0.47 lbs./ft.
3 ½"	2.50"	0.66 lbs./ft.

Types of Poly Pipes Designed for Any Particular Well:

- Seamless HDPE liner (with a temperature range of up to 160°F)
- Seamless polyolefin liners (with a temperature range of up to 210°F)
- Seamless thermoplastic liner (with a temperature range of up to 300°F)
- Seamless thermoplastic liner extreme tube (with a high temperature range)

Polybutylene is basically used in industrial and housing water pipe connections and distribution. Polybutylene can be used for low-pressure valves and fittings for plumbing.

Section II

Principal Application of Tubing anchor catcher (TAC)

Do You Know What a TAC Is For?

The TAC is a retrievable tool. It is designed to anchor the tubing in the casing string, maintain tension on the tubing inside the casing string, and catch the tubing string in case it becomes parted. The TAC will keep the tubing string from breathing and minimize the pipe moving from side to side or up and down during rod reciprocations.

The TAC is a mechanical set tool/device that is run on production tubing to anchor the tubing string in tension inside the casing. The built-on mechanical slips/dies enable the tool to hold bites and maintain a tight tension grip against the casing wall when it is set properly. The purpose of holding tension on the tubing string in a well is to reduce tubing and sucker rod movements during sucker rod reciprocations (see-saw movements). The reduction of pipe movements and tubing buckling will improve pump efficiency, reduce rod wear, and may prevent holes in the tubing, parted pipes, and friction holes in the casing string.

The Absence of a Tubing Anchor in Beam Pumping

The movements of a small-diameter tubing string inside a larger casing will be considerably high and frequent without a tubing anchor.

- Tubing buckling (caused by rod movements)
- Uplifting motions (fluid buoyancy and upstroke movements)
- Downward motion and friction (rod movements)
- Side-to-side movement (rods and pump strokes)

An unstable tubing landing will cause repeated rod parts, tubing holes, and a reduction in production. The TAC should be landed in tension below the pump seating nipple to be effective (set the anchor across a bonded casing area if possible).

The TAC is designed to be set in tension, not compression. Make sure you do not run a TAC upside down (it will be difficult to fish out). A tension-set TAC will reduce tubing buckling and rod wear during reciprocation. A tension-set tubing string may maximize pump efficiency and may reduce parted rods and holes in the tubing string. Losing the bow drag springs on the TAC is a major problem in all H2S wells (leaving junk blades in the well that are difficult to fish out).

Major Functions of TAC

- Set and anchor tubing with tension in the casing string
- Prevent tubing string from falling in case of tubing becoming parted
- Reduce tubing buckling and breathing during rod movements
- Increase pump efficiency and increase production
- May reduce rod and tubing wear in deviated and dogleg wellbores

For your information, there is no such thing as straight hole drilling in the oil field (only controlled drilling). Three or four degrees from the vertical line is the best way to drill straight holes in the oil and gas wells of today. In oil well operations, we fast-drill the holes and worry about paying the price for it later during the production operation. This is why we have so many tubing and rod failures (it may not have anything to do with chemical treatments).

The TAC is a tension and compression set tool that may provide efficiency in beam sucker rod application and PCP artificial lifting.

The TAC may be used as an anchor for electric submergible pumping operations to prevent loosing parted tubing with an electric cable down the hole. For a tubing anchor to be effective, it must be set properly below the pump seating nipple.

❖

Operating a TAC

Running, Setting, and Releasing Techniques

Note: Serious injury can occur while setting and releasing a tubing anchor.

TAC Running Procedure

- Inspect the TAC on the rig floor. Check the pin and collar threads to ensure they are in good standing.
- Check and inspect the bow springs (drag springs). Use double springs if needed.
- Apply a fresh and clean thread compound (pipe dope) on the pin end and collars properly.
- Function-test the tools by rotating the body to the running position, clean the tool with diesel, and lubricate the components using lubricate oil or grease.
- Make up the TAC on the tubing string as required (I prefer to install the TAC below the seating nipple, below the pump, if appropriate).

Tubing and TAC Landing

Run in the hole as listed below (from the bottom up):

a) Blind bull plug
b) Mud anchor joints (minimum of two joints)
c) 4' or 8' perforated nipple
d) TAC
e) 8' spacing pup joint

f) API seating nipple with 20' gas anchor

g) Production tubing string to the surface

Before running the tool, hand-turn the TAC on the rig floor first to make sure the slips on the tool function properly and the slips are engaged properly before running it into the well.

a) Check the slips to ensure they are sharp enough to grab/bite in the casing.

b) Check the shear pins and make a record of them.

c) Check the drag springs to ensure they are tight without cracks.

(The drag springs on the TAC are of inferior quality and need improvement. They are a major problem in H2S wells. Most of the wells in West Texas and New Mexico are full of broken drag springs. H2S will penetrate the springs at the bending point.)

- Check the ratings of the shear pins inserted in the TAC (10K, 20K, 30K, 40K or 45K). Each of the shear pins on the TAC is rated at 5,000 lbs. (made with brass pins). The purpose of the shear pins on the tubing anchor is to shear and get the tubing string free in case of a problem when releasing the tubing anchor.
- Make sure you are comfortable with the shear pin ratings on the tool in case of emergency to release or shear and pull the tubing string out safely.
- Make sure your tubing string is rated for tension and has a higher shear value to release the TAC in case of emergency.
- The tensile strength of the tubing and the TAC rating must be evaluated in all the deeper wells. For example, a TAC set at 10,000' on 2 ⅞" (6.50#) tubing = (65000# in air) plus a 40,000 lb. shear valve on the TAC (65,000 plus 40,000 plus the weight of the blocks, which is 4,000 lbs.).
- Check whether your tubing string is good enough to withstand the pulling weight (make sure to use low-profile slips to prevent over-pulling into the TAC).
- Always zero the indicator on the rig before picking up on the tubing string.
- Make sure your drill line is in good standing to avoid accidents.
- The crew must clear the rig floor before pulling onto the tubing string.

Caution:

- Do not run and set a TAC in bad or old casings with holes.
- Do not run more than five joints of tubing below any TAC.
- Do not run a TAC in a well with high H2S gas concentration.
- Do not run a TAC with flat or dull slips.
- Do not run a TAC below open perforations (high risk of fishing work).
- Use double drag springs if necessary. The purpose of drag springs is to keep the tool stationary against the casing while turning the pipe to set and/or to release the TAC (bow springs are used for friction).

- If you lose the drag springs in the well, you may or may not be able to set and/or release the tubing anchor. Drag springs act as drag blocks against the casing string. There are three single drag springs on each TAC. I recommend using tandem drag springs screwed on the TAC in deeper and non-corrosive wells (will explain to you later).
- Trip in the well with the TAC slowly to the running depth as required.
- Check the tubing string for the correct landing depth.
- Production packers and tubing anchors should be run and set in a casing spot with good cement bonding outside of the casing. Keep away from casing collars.
- Always check the well for flow first before removing the blowout preventer (BOP) stack.
- Keep good pipe measurements and space them out before setting the TAC.

Motions to Set Standard TAC

- Check the tubing tally for the correct landing depth.
- Zero the indicator on the rig for accurate weight reading.
- Pick up on the tubing string and read the tubing weight on the indicator.
- Mark and space out as needed.
- At the desired landing depth, rotate the tubing string six to eight rounds to the left (counterclockwise) with a pipe wrench for the anchor slips to bite and catch inside of the casing wall.
- After the slips make contact with the casing, slowly pull with the desired level of tension onto the tubing string while you still have left-hand torque on the pipe.
- Work the pipe up and down several times to relax and to set the slips.
- Release all the left-hand torque on the tubing and make the final tension as required.
- Repeating the up and down motions will relax the landing slips on the TAC. Release any torque that is left on the pipe before landing.
- Make sure the tubing joints are properly made up to prevent the tubing from backing off while setting the tubing anchor using left-hand torque.
- Mark the pipe and install wellhead hanger slips (the hanger slips must be flat). Do not use hammer blows to drive down or push the slips in place.
- Nipple up the tubing wellhead and equipment.
- Do not over-pull on the tubing string and TAC. You may damage the old casing, or you may part the tubing string.
- Always keep a good report and information on the tubing landing and TAC (tubing weight, tension, and shear pin valves are important).

Note: The landing equipment on the wellhead could be a slip-type and/or screw-type threaded flanged tubing head.

- If the landing equipment on the wellhead is a slip type, such as a Larkin head

and/or a similar brand slip-type tubing landing hanger, it can be fairly simple and easy to set and/or release the TAC.

- If the landing equipment on the wellhead is a screw-type flange, you are subject to lose some of the tension that you pulled against the TAC when slacking off to nipple up the flange.
- All the TAC tension landing on the screwed flange wellheads must be carried out using low-profiled slips to prevent you from pulling too much tension into the anchor slips and casing string.
- Do not use standard rig slips on screw-type tubing flange heads on the well. Standard rig slips are too tall and will tear the casing, part the tubing string, and/or shear off the pins on the TAC. (It has happened before, and the well owner will not have any idea what happened and what caused the casing damage.)

Estimated Tension Values for TAC Landing

Tubing Size	Depth	5 pts.	10 pts.	15 pts.	20 pts.	25 pts.
2 ⅜", 4.70#, 4.70#, J55	2,000'	3"	6"	9"	12"	15"
	3,000'	5"	9"	14"	18"	23"
	4,000'	7"	12"	18"	25"	30"
	5,000'	8"	15"	23"	31"	32"
	6,000'	9"	18"	29"	37"	42"
	7,000'	11"	21"	32"	42"	51"
	8,000'	12"	25"	37"	48"	60"

	9,000'	14"	28"	41"	52"	60"
	10,000'	15"	31"	45"	60"	70"

****

	Depth	5 pts.	10 pts.	15 pts.	20 pts.	25 pts.
2 ⅞", 8ʳᵈ	2,000'	2"	4"	7"	9"	11"
6.50#, J55	3,000'	3"	7"	10"	13"	16"
	4,000'	4"	9"	13"	18"	21"
	5,000'	6"	11"	17"	22"	28"
	6,000'	7"	13"	20"	21"	30"
	7,000'	8"	15"	23"	31"	32"
	8,000'	9"	18"	26"	35"	41"
	9,000'	10"	20"	30"	40"	55"
	10,000'	11"	22"	33"	41"	52"

****

	Depth	5 pts.	10 pts.	15 pts.	20 pts.	25 pts.
3 ½", 9.5#/ft	2,000'	1"	3"	4"	6"	7"
J-55, 8ʳᵈ, T&C	3,000'	2"	4"	7"	9"	10"
	4,000'	3"	6"	9"	12"	14"
	5,000'	4"	7"	11"	14"	18"
	6,000'	4"	9"	13"	18"	21"
	7,000'	5"	10"	16"	20"	25"
	8,000'	6"	12"	16"	23"	28"
	9,000'	7"	13"	20"	25"	33"

TAC Releasing Techniques

Do not fight with the tools; work with the tools. Problems with releasing tubing anchors could give you expensive fishing work and/or cause you to lose your wellbore. Releasing the TAC may cause accidents and/or serious injury. Do not allow the rig crew to set or release any mechanical set tools without direct supervision. Trained, knowledgeable, and experienced personnel are required to successfully run, set, and retrieve the production tools in or out of a well safely.

Releasing and Retrieving the TAC

Rig up on the well (ensure good-condition drill line and tubing string). Circulate to kill the well and also to wash and flush out any possible solid deposits around the TAC if possible (solid buildup around the TAC is normal). If you have a shallow hole in the tubing string, you may not be able to successfully circulate the well at the bottom. (You may be able to bullhead the kill fluid into the perforations only.)

There are several important downhole factors that you may need to know before releasing tension-set equipment out of a well:

a) Tubing size and condition (how long the tool has been in the well)
b) How much tension is left on the tool and tubing (TAC and/or tension packers)
c) Shear value on the TAC (30,000# or 45,000#) plus the tubing weight and the weights of the rig blocks (always zero the weight indicator to get accurate readings)
d) Condition of drag springs on the TAC because of the CO2/H2S environment

Make sure the well is dead before removing the wellhead equipment. The wellbore fluid must be in static condition during the entire workover operation. Never assume a well is without a H2S gas flow.

Wellhead equipment on beam pumping could be a slip-type and/or screw-type flanged tubing hanger (slip-type hangers could be Larkin, Hercules, or Hinderliter style). Note that hanger slips in Larkin, Hercules, and Hinderliter tubing heads are designed to hold the tubing string in tension and prevent the pipe from falling into the well. High wellbore pressure can push the tubing string on an upward lifting motion and out of the slip (Durkee field salt water disposal).

If the tubing hanger is a Larkin-type slip, simply pick up on the tubing string a few inches upward and remove the slips from the wellhead (do not lose the slip segments in the hole). Use a split tubing wiper to prevent any part of the slip from falling back into the well. If you drop any of the slip segments in the well, it will be difficult to pull the TAC out of the hole.

If the hanger is a screwed flange type, do not use the tall profile rig slips to lift and remove the tubing head flange. This may cause excessive tension strain into the tubing string, causing damage to the casing, and may part the tubing string or shear off the anchor. When using standard tall profile rig slips, you may have to pull sixteen inches upstream into the tubing string and the tubing anchor plus the weight of the tubing string to remove the wellhead flange.

Use low-profile tubing slips to set and remove the tubing bonnet with ease without pulling unexpected tension into the tubing string safely. (Always zero the weight indicator and check the pipe slips' ratings before application). Low-profile slips are only six inches tall, and standard rig slips are fourteen inches tall **(make sure you check on any pipe slip weight ratings before application to avoid accidents or losing objects in the well).**

Check the well for flow before backing off studs and nuts on the tubing head. Install a short pup joint onto the wellhead and pick up on the tubing flange. Back off and remove all the studs and nuts. The crew must stand clear when pulling onto the tubing and TAC assembly (anything can go wrong). Pick up on the tubing head equipment enough to clear the flange and to set the slips. (Clear your body and fingers from the flange while it is in tension.)

Back off and remove the flange using rig blocks. Install a pup joint onto the tubing string. Pick up on the tubing string and pull the slips out safely. Slack off with the tubing string to a neutral position and nipple up the BOPs. Rig up the floor and handling tools

as directed. You are ready to release the TAC now.

How to Release TAC

Avoid near accidents or injury. To release a TAC, slack off on the tubing string to find the neutral tubing weight first. Slack off one point weight on the TAC and the tubing. Work the tubing string slowly up and down to fill for the tubing condition and TAC movements. Never jar down on the TAC (do not fight with the tools).

With slight compression on the tubing string, turn the tubing string to the right for five to eight rounds at the tool to retract the cones and also allow the slips to move into the tool housing. Use a hand pipe wrench or as directed. Move the tubing up and down several feet while turning the pipe to the right to release. You will see the tubing losing torque, and the TAC will become free (or often jump free). Continue turning the pipe a few extra rounds to free the anchor slips and to relax the upper and lower cons on the TAC.

Start on an upward motion, pulling the tubing string slowly. Use backups to keep the tubing string from turning left. Any left-hand turns may reset the anchor and may cause the tool to set while tripping the pipe (this may cause accidents, part the tubing, or shear the tool). If normally releasing the TAC is impossible, the tool may be sheared off before pulling the tubing string. (Consult your supervisor before pulling into the tubing to shear the TAC.)

High-tension load on the tubing string is required to shear off the TAC. Check the rig and lines before attempting to shear off the anchor (parting the tubing string is highly possible). If the TAC is not completely released, slack off and rotate the pipe a few additional rounds using a hand wrench to get the anchor free (do not use tongs to release the tool, if possible). Note that too many right-hand turns may damage the tubing anchor.

Do not drag out any partially released TAC or isolation packer up the hole (it may be hung up and/or shear off, causing accidents or fishing work). A sheared-off TAC will often be slow and difficult to pull out of the hole because of loose slips. Loose slips may be hung up at a shallow depth, and you may not be able to pull up or go down.

You may become stuck. A stuck anchor at a shallow depth is difficult to pull or mill out (a stuck fishing anchor at a shallow depth is costly). You may have to mill out the slip segments to push the anchor down the hole. Do not fight with the tools; work with tools to avoid fishing accidents or injury (get help when needed). Turn the pipe to the right for two rounds every twenty stands while pulling the tubing out of the hole; this may prevent the tubing anchor from resetting and being hung up, causing accidents. Note that on progressive cavity artificial lifting pumps (PCP),

the TAC has <u>right-hand–set</u> rotations and <u>left-hand–release</u> rotations. If you decide to release the TAC with a pipe wrench, you may need two rough necks to unset the tool.

When you are turning the pipe with a pipe wrench, you will feel lots of torque on the pipe wrench (watch for tripping or falling hazards on sharp objects). Hold onto the pipe wrench with torque and do not let go. Loosening the pipe wrench may cause it to backlash and hit you on the side, causing serious injury. When the TAC becomes free, the tubing string may spin or jump free. Your pipe wrench may lose torque and cause you to fall. When a TAC stays in a well for several years, it will be difficult to release. Floating formation solids and scales will build up onto the anchor slips and cause cementation around the tool.

If you are unable to release the TAC by hand, consult your supervisor. You may rig up the floor and handling tools to use tubing tongs as appropriate. Get work direction from your supervisor (watch to make sure you are not over torque and twisting the tubing while repeatedly turning the pipe). Do not over-torque the pipe, causing it to twist off the tubing string. This will cause a long and complicated fishing job, and you may be there fishing the tubing for some time and/or T/A the well (you are subject to losing the well).

In H2S-environment (especially West Texas and New Mexico) oil wells, you may expect the TAC to come out of the well without drag springs (wellbore full of junk blades). H2S will penetrate the bending point of the anchor and create corrosion damage (hairline). Most of the wells in West Texas and New Mexico are full of lost steel bow springs that have not been reported or ignored. Lost drag springs are major obstructions in the wellbore, especially in deviated and horizontal wells; junk blades are tough to mill and fish out. That is why you may not be able to clean up a horizontal well to the end.

> ➤ There are several TACs in the market today.
> ➤ Some tubing anchors are designed to set and/or release only two rounds.
> ➤ Some are used without drag bow springs.(set and release with pressure)

Do not run a TAC in a well with heavy paraffin,

unconsolidated formation sand, or a bad casing string (good luck). Note that releasing a TAC with several joints of tail pipe below the anchor could cause a complicated problem in sandy wells. If the tail pipe below the anchor is stuck in formation solids, it will be difficult to transfer torque to release the tubing anchor (hand-releasing will often be difficult and risky). Using pipe tongs may cause a back-off swing on the tubing string below the anchor and may part the tubing string (similar to electric string shot).

Advantages of TAC

a) Prevents upward and downward movements caused by sucker rods
b) Increases pump efficiency
c) Reduces friction and wear on tubing couplings and casing

d) Minimizes tubing thread leaks

e) Reduces rod repair

Disadvantages of TAC

a) Spring are a major problem in a H2S environment
b) Shears off on the tool will slow down the operation and may cause fishing

c) Difficult to release in a crooked and sandy wellbore and may cause injury
d) Difficult to mill out (wish you did not have it in the well)
e) Causes formation buildup around the tool

f) Chokes and blocks off the gas flow up the annulus
g) May cause damage to casing string

Important Tools and Equipment Used on Beam Pump Artificial Lifting

A. Stuffing Box on Beam Pumping

The purpose of the stuffing box is to contain well pressure, prevent the fluid from escaping, wipe off the polished rod, and prevent oil/water pollution. The stuffing box

is a device packed with rubber cups or rubber rings to pack off around the polished rod tightly, prevent oil, water, and gas from escaping under pressure, prevent fluid leaks, and avoid surface ground pollution.

The stuffing box's rubber elements must be lubricated properly and changed in a timely manner to avoid pollution. The rubber elements must fit in the stuffing box and the polished rod and/or polished rod liner size. Dry stuffing box elements will fail and cause the fluid to escape.

There are several types of stuffing boxes in the market to choose from. A poorly maintained stuffing box is one of the major sources of oil, water, and gas leaks on pumping wells in the oil field. The stuffing box rubber elements may be of different shapes and sizes based on the stuffing box manufacturing design. The rubber elements must be of correct size to fit tightly around the polished rod and/or polished rod liner.(damaged packing will created liner holes)

The polished rod and polished rod liner

will have different diameter sizes. If you are using polished rod–sized packing elements in a stuffing box with the polished rod liner,

you are using the wrong packing element; it will be too small for packing and will prevent the polished rod liner from going down. In this case, the unit will throw the bridle out on the horse head and may bend the polished rod. It will cost you to bring a workover rig to correct your mistake. (Pay attention when installing the stuffing box elements.)

Overtightening the pack-off rubbers in the stuffing box will burn and damage the rubber elements, create a large hole in the rubber pack-off, and will cause oil, water, and gas to escape. Torn rubber pack-off elements may fall down the hole and cause the rod string to part. A lack of lubrication and overtightening the packing elements will create heat and will burn the packing elements. Dry and tight packing elements will cause friction and may create fires.

The polished rod must be aligned vertically and straight at the center of the stuffing box

to avoid tearing and wearing the stuffing box elements on one side. The pumping unit head (horse head) must be straight and vertical to keep the polished rod **straight** and **level** at the center of the stuffing box. A worn center bearing on the pumping unit may often cause shifting or push the polished rod off to one side, causing damages. Lubricating the stuffing box will make the polished rod cool and prevent the packing elements from leaking.

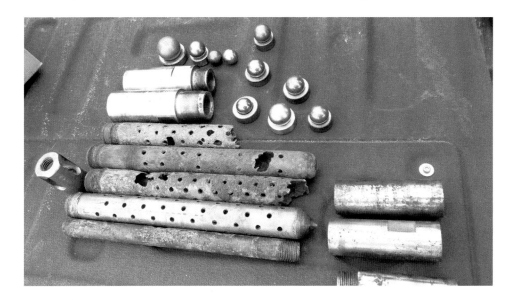

Dirty well fluids such as high-cut salt water, sand, mud, and iron sulfate will damage the packing elements (causing more leaks and requiring more packing changes). Salt water is not a good lubricant for the stuffing box elements. One of the main sources of stuffing box lubrication is a clean well fluid (produced oil and clean water will lubricate the packing and cool off the polished rod assembly, which will prolong the life of the packing elements).

A hot polished rod is an indicator of downhole mechanical problems such as holes in the tubing

string, parted rods, or overtightened packing elements or pumping off the well. Reciprocating a hot polished rod will burn the packing elements, will cause oil and gas to escape, break the polished rod, and may create fires. Unleveled stuffing box with cause liner damage!

Use a lubricating box (oiler) to lubricate and cool off the polished rod and packing elements on all low-producing pumping wells. Apply oil on all low-fluid or high-gas cut pumping wells. Parted rods and holes in the tubing string will cause the polished rod to become hot and burn the packing elements. Never hold a hot polished rod while in motion. Use clocks on all low-producing stripper oil and gas wells.

Caution:

- Do not operate a pumping unit on a high-pressure flowing well. A high gas/oil ratio fluid flow may prevent the subsurface fluid pump from operating properly (high pressure will prevent the valves from operating properly).
- Operating a pumping unit on a high-pressure flowing well may cause stuffing box failure (do not operate a pump unit on a flowing well with a surface pressure of more than 300 psig).
- High-pressure gas may dry off, cut, and burn the packing elements in the stuffing box quickly and may cause an uncontrolled fluid flow out of the stuffing box. It may create pollution and cause complicated well control issues. (Let's hope you will have a good-condition BOP below the stuffing box.)
- Never use an isolation packer in a flowing well that is on artificial beam pumping.
- A flowing well with high well pressure on beam pumping may burn the packing and be uncontrollable. It will be extremely difficult to kill the well and stop the oil pollution.
- An isolation packer on beam pumping will prevent you from killing the well through the annulus.
- If you perforate a new reservoir in a well with beam pumping/sucker rod installations, do not put the well on a beam-rod pump (it is a live well).
- Use an isolation packer and swab the fluid first to allow the well to flow naturally for as long as possible before putting the well on beam pumping.
- You may lay down the rods individually. Protect and safeguard your rods and the subsurface pump from corrosion while flowing the well.
- Continue to flow the well until the flowing pressure declines below an acceptable shut-in pressure of 100 psi or less before using an artificial beam pumping unit to lift the fluid.

B. APPLICATION OF PUMPING TEE (FLUID TEE)

A pumping tee is three-way T-shaped extra-heavy-duty carbon steel metal, available in different sizes. The pumping tee has extensive use in beam pumping operations. It is screwed on top of the production tubing string above the ground or on top of the wellhead flanged assembly as a fluid outlet. It may be installed above the BOP and/or below the stuffing box above the ground.

I prefer to install an ex-heavy pumping tee in the order below:

a) Stuffing box (at the top)
b) Pumping tee (in the middle, on top of the BOP)
c) BOP (on the top of the wellhead or tubing string)

See the attached wellhead picture with a pumping tee. The pumping pee is threaded on all three sides to accommodate the fluid flow direction. It is available in different sizes and uses a dependable IPC ex-heavy material.

C. BOP ON BEAM PUMPING

This device is screwed above the wellhead flange and below the pumping tee and the stuffing box. (Some operators may install the BOP equipment differently) BOPs are designed to pack off around the polished rod and/or the polished rod liner in case of pressure emergency only. The BOPs must stay fully open during the pumping unit operations (for emergency use only)

There are two sets of specially made heavy-duty rubber elements inside of BOPs to fit around the polished rod and/or polished rod liner. BOPs are necessary to isolate the well flow in case of emergency and/or change the packing elements in the stuffing box in case of any fluid leaks above the ground.

In low-producing oil wells (strippers wells), you may find no BOP at all (just a pumping tee and a poor-boy stuffing box with rubber elements). The rubber elements in the BOP must be of correct size, undamaged, and free of sand and should fit tightly around the polished rod or polished rod liner to avoid hazard gas from escaping when an emergency occurs (safety and pollution issue). Check the pressure rating on all of the wellhead equipment before installation. Use high-quality, high-pressure equipment above the wellhead. The BOPs may be designed for specific rated applications.

D. POLISHED ROD (CHROMED AND POLISHED)

The piston polished rod is made from a cold drawn carbon steel alloy with high tensile strength to lift the dynamic loads. The polished rod is either chrome plated from end to end or without chrome at both ends based on the pump stroke (a non–chrome-polished rod must be spaced out correctly to prevent the stuffing box from fluid leaks). The application of polished rods is based on wellbore specifications to avoid rod failure (corrosion, abrasion, and load).

The polished rod is the top connection on the sucker rod string. It is a solid heavy-duty bar that is honed and polished, chrome plated, and slick from the outside. Polished rods are available in different sizes and lengths. The polished rod is connected to the sucker rod string below the wellhead, extends upward through the pumping tee, BOP, and stuffing box, and is held securely by the polished rod clamp, carrier bar, and wire rope bridle at the horse head above the ground. The polished rod is the only bar that holds the entire weight of the rod string with dynamic load while reciprocating and/or standing still. (Polished rods may be used with or without liners.)

Polished rod liners are hollow tubes honed and polished from the outside. The liner must be tightly secured with available bolts and rubber rings to the polished rod to avoid fluid leaks (leaks between the liner and the outside of the polished rod). Any prolonged high-pressure leaks will split the liner and cause pollution. Pack off friction will cause holes in polished rod liner

The polished rod may be used with and/or without a polished rod liner on some rod string designs. The purpose of using a polished rod liner is to protect the polished rod only. Pull, clean up, and lubricate the liners before installation to keep them from becoming frozen and stuck onto the polished rod because of internal corrosion and fine solids. Avoid liners to drag on one side

E. POLISHED ROD CLAMP

A polished rod clamp is a device that is made up and mounted on the polished rod and held by a carrier bar at the end of the wire rope. Clamps are available with one-, two-, or three-bolt configurations based on the application. The polished rod clamp is a heavy-duty bolted device made up on the polished rod to keep the entire rod string from falling or sliding down the hole. Polished rods with three-bolt clamps will be used in deeper wellbores (8,000' to 12,000').

Polished rod clamp/clamps are set and held on the polished rod after spacing the rods. The polished rod clamp is set and landed against the carrier bar, which is connected to the horse head wire rope bridle. The polished rod carrier bar must be locked in place with a special door to keep the polished rod from jumping out and dropping onto the stuffing box, causing damage at the surface and downhole.

Do not use a cheater pipe when tightening the polished rod clamp on the polished rod. You may cause hairline cracks or damages to the polished rod, causing the polished rod to break and become parted just below the clamp. In deeper wells, you may use two polished rod clamps if necessary.

Care and Maintenance of Polished Rod

Piston polished rods are basically free of maintenance if properly installed.

- The polished rod must be straight and vertical without restrictions.
- The polished rod is chrome plated on the surface (it is polished, slick, shiny, and clean). It must be kept clean, lubricated, and without cracks, dents, bends, scratches, or grooves to function properly and prevent leaks.
- Avoid hammer blows on any part of the polished rod or polished rod coupling.

- Always back off the polished rod coupling using friction wrenches only.
- Do not overtighten the polished rod clamp/clamps (no cheater pipe).
- Always use a polished rod box

on the polished rod. The polished rod box

has more threads than regular standard rod boxes.

- Over-torqueing the polished rod clamp against the polished rod will cause hairline cracks and damage the polished rod surface.
- Use three-bolt clamps and/or two clamp devices for better grip (no cheater pipe).
- Always hang the polished rod straight, with the carrier bar flat and level.
- Make sure the carrier bar is held flat and level to avoid forcing the polished rod into a bending position (avoid stress and strain positions).
- Change the stuffing box packing more often to avoid damage to the polished rod.
- Do not overtighten the stuffing box packing elements against the polished rod.
- Use an **oiler** to lubricate and cool off the polished rod. Use lube oil only.
- Only two feet of the polished rod should be sticking above the polished rod clamp.
- The polished rod is mostly designed to be below the wellhead and not sticking high above the horse head. Otherwise, you are wasting the polished rod.

- The horse head and the pumping unit must be aligned at the center of the well to avoid damaging the polished rod and the stuffing box's packing elements.
- The polished rod must be straight at the center of the horse head groove and at the center of the stuffing box while reciprocating (not leaning to one side).
- Shut down the pumping unit if the polished rod temperature becomes too high (evaluate the subsurface and wellbore for downhole problems).
- Lubricate the stuffing box using grease and lube oil only.(grease may plug off back pressure)
- Avoid jarring, water hammering, and pounding while reciprocating.
- Clean up and remove formation scales as well as salt and iron sulfide build-up inside the stuffing box and around the polished rod (abrasion material).

- Do not use a bent and dented polished rod on the rod string

Section III

Subsurface Conventional Fluid Pumps

Downhole Fluid Pumps

Subsurface fluid pumps are basically reciprocating mechanical type pumps. They consist of a hollow polished pump plunger

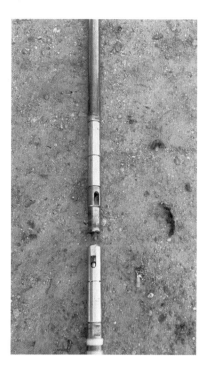

and a cylindrical hollow polished barrel with a built-in standing valve

(inlet valve) and traveling valve (discharge valve).

There are basically two types of subsurface conventional fluid pumps:

A) Insert pumps (often called rod pumps):

There are several types of insert pumps to select. The term "insert pump" is derived from the fact the all the components of the pump, such as the valves and plunger, are inserted or built inside of a closed pump barrel.

B) Tubing pumps: The name "tubing pump" is derived from the fact that the pump barrel and seating nipple are made up on the end of tubing and become part of the tubing string. The tubing pump consists of three **separate** components that are run on the tubing and the sucker rod string.

The basic components of any subsurface fluid pump are as follows:

- Pump barrel (a hollow tube threaded on both ends—internal and external)
- Pump plunger (polished heavy-duty tube designed in different shapes)

- Traveling valve (ball and seat)/discharge valve

- Standing valve (ball and seat)/inlet check valve
- Seating nipple mechanism

The machined seating nipple or standard API seating nipple/seating shoe is designed to Isolated the fluid and hold the standing valve stationary in place.

What Is the Insert Pump (Rod Pump)?

(Read the content carefully to understand pump types and their applications.)

On the insert fluid pump, all the components of the fluid pump are made up and held together as a complete unit. All the components of an insert pump are made up inside of a polished barrel before it is run into a well. Insert pumps can be pulled for repair and replaced without pulling the tubing string. The insert pump is made up with a single polished barrel with the valves and plunger. It is built in as a closed tool (you will not see the plunger and/or the internal components of the insert pump by looking at the pump from the outside).

Critical Components of a Subsurface Fluid Pump

[Show a picture of a pump with its components.]

A. Pump Barrel

Pump barrels are made from special hollow brass/bronze/steel tubes based on the pump design and application requirements. There are basically two types of manufactured pump barrels:

- **One-piece** heavy-wall seamless pump barrel (H)
- Thin-wall seamless pump barrel (W or S)

Pump barrels come in various lengths and sizes based on the wellbore requirements. They are internally treated with a special chrome alloy coating to resist corrosion and abrasion. They are made with .002 to .005 inches of tolerance between the barrel and the selected plunger.

B. PUMP PLUNGER

An ex-heavy wall carbon steel bar with opening hole passing through from end to end. The most popular plungers are the following:

1) "Plain" pump plunger
2) "Grooved" pump plunger

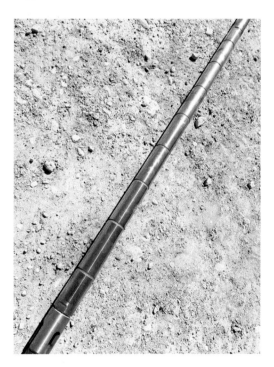

3) "Soft ring" packing plunger (for shallow to mid-range deep wells)

In deeper wellbores, closer fitted metal-to-metal plungers should be used to reduce fluid slippage and improve pump efficiency. All the pump plungers are hollow, honed, and chrome plated, with accurate space tolerance to fit inside a pump barrel. Pump plungers are made to resist against severe wellbore corrosion and abrasion problems. The plunger is selected based on the barrel length, stroke length, and wellbore conditions.

Plungers may come in different configurations and sizes:

- Pin-end spray metal plunger
- Box-end spray metal plunger (thinner wall)
- Box-end spray grooved plunger
- Monel plain metal plunger
- Soft-packed ring plunger
- No-lock plunger

The selection and application of any of the above plungers is based on the physical condition of the wellbore (friction, sand, gas, and fluid viscosity). **There is not such thing as an ideal wellbore for pumping operations**. The selection of the pump type and its components may be different from one well to another.

Application of "Plain" Metal Plunger

Plain metal plungers are tough for deeper wells. Sand, scale, and abrasion particles may get trapped around the plunger and may cause severe friction scores along the length of the plunger and the barrel. A stuck plunger will cause parted rods and incur pulling costs (clean wellbore fluid is recommended). Plain metal plungers have lower percentages of fluid slippage than grooved plungers.

Application of Grooved Plunger

Grooved plungers may last longer in a well with some sand particles. The grooves on the plunger may have greater advantage in reducing sand and solids that pass between the plunger and the barrel. Sand grains will cause scores and deep abrasion on the face of the plunger and the barrel and will cause pump failure. Sand grains will be trapped or lodged inside the plunger grooves and may prevent solid bridges between the barrel and the plunger. Grooved plungers will have more fluid slippage than plain-type plungers. The application of too many grooves on the plunger will make the tools less efficient.

Soft-Pack Pump Plunger

The soft-pack pump plunger is designed for shallow to mid-range well depth. Soft-pack plungers have good resistance against corrosion and sand abrasion. Plungers are designed with clearance between the metal pump plunger and the ID of a pump barrel. The clearance gap may range from 0.001″ to 0.005″ (0.001″, 0.002″, 0.003″, and 0.005″) considering the fluid viscosity in plunger selection. Note that all the plungers will have slippage loss (based on the gap between the plunger and the barrel).

Subsurface Fluid Pump Valves (Balls and Seats)

There are two types of valves in a fluid pump: **flat and ribbed**.

a) Traveling Valve (Ball and Seat): Traveling valves are either closed-type or open-type cages. The traveling valve is the fluid pump discharge valve that acts as a check valve to retain fluid in the tubing string during the upstroke motions. Traveling valves may be made with a single ball and seat or double balls and seats.

b) Standing Valve (Ball and Seat): The standing valve is basically the pump's intake valve. It acts as a check valve by opening and closing, allowing fluid into the pump chamber. As the name implies, the standing valve is a stationary valve in the pump. Closed-cage standing valves are generally used in tubing pumps.

On the insert pump, all the parts are made up together before it is run into a well as one piece and appears as a closed assembly (you cannot see the inner parts after it is made up). Insert pumps have a wide range of applications in deep oil and shallow oil wells (ranging from seven hundred feet to twelve thousand feet). Do not run a brass/bronze insert pump in a well that is acidized (acid must be swabbed out first).

Seating Element Configuration

All artificial rod pumps require seating configurations. Seating nipples are run on the tubing string so that the fluid pump can be run and anchored securely inside the tubing string.

There are two types of seating mechanisms:

a) Nylon cups and/or composition rings, larger than seating nipple IDs, are inserted on the pump and used as seating assemblies. The purpose of the seating mechanism is to create a tight seat inside of the seating nipple to hold the pump in position and avoid fluid leaks.

b) The mechanical hold-down seating assembly consists of a cone-shaped brass/bronze soft-seal metallic element.

Fluid pumps are selected and designed based on the following:

- Pumping unit size and stroke length
- Volume of fluid to displace
- Wellbore fluid characteristics (viscosity, corrosion, abrasion)
- Wellbore static fluid level and volume
- Solid contents (sand, mud, salt, scale)
- Gas and liquid content (oil, water, and gas ratios)
- Well's pumping depth (depth of seating nipple landing)
- Downhole mechanical conditions (casing, tubing, rod conditions)
- Tubing string size and grade (bare tubing or IPC)
- Fluid corrosion content (content of H2S and CO2)
- Gas/liquid ratio
- Viscosity of oil
- Volume of fluid to produce

The application of a pump-type design must be discussed with the pump manufacturer to ensure the correct pump selection and pump size before it is run into a well. The change of oil/water ratio and viscosity of the wellbore fluid must be considered.

Classification of Rod Pumps (Insert Fluid Pumps)

There are basically two types of insert pumps:

1) Bottom hold-down insert pumps

2) Top hold-down insert pumps

❖

Bottom Hold-Down Insert Pumps

There are basically two types of bottom hold-down insert pumps:

a) Mechanical (can be bottom hold-down or top hold-down) (stationary pump
b) Standard cup (stationary and travel barrels with extensive use in deep or shallow wells)

A. Mechanical Bottom Hold-Down Insert Pump

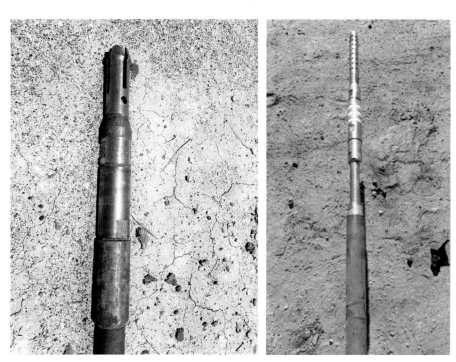

For the mechanical bottom hold-down insert pump, the pump barrel will stay stationary, while the plunger travels up and down with sucker rods. The mechanical pump hold-down has a build similar to that of the standard cup-type pump with two distinct and noticeable exceptions:

1) Special mechanical lock sealing element (no nylon cups)
2) Special type of mechanical seating nipple

   a) Bottom lockdown seating nipple
   b) Top lockdown seating nipple

➤ On the bottom of the mechanical hold-down pump,

the hold-down mechanical latching device is made up at the bottom of the standing valve.

➤ The mechanical hold-down pump is run and snap-latched into a special mechanical seating nipple that is made up and run on the tubing string (nothing else can be added below any mechanical hold-down locking assembly).

➤ The mechanical sealing part of the tool is made from soft brass/bronze to latch, seat, and seal off the inside of the special mechanical seating nipple, just like cup-type sealing elements (no nylon seating cups).

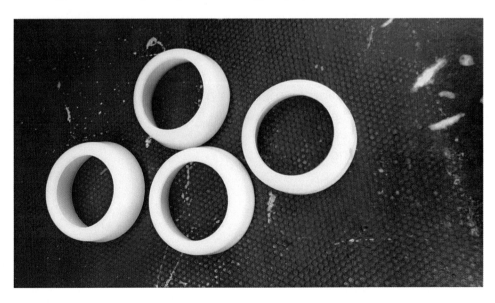

➤ No additional tools can be screwed below the bottom hold-down assembly.

> ➢ On the mechanical hold-down pump,

the seating nipple and seating elements are different from those of the standard insert pump. (Do not worry; it will work fine and will not leak if you seat the pump correctly.)

[See the picture.]

The mechanical seating nipple

is built differently from the standard API seating nipple. The mechanical seating nipple is shorter in length (7.4"), threaded internally with a machined stand-off at the bottom end, and externally threaded on both ends of the seating nipple (do not run the seating nipple upside down).

The mechanical seating nipple has a built-in machined stand-off groove near the bottom, just above the female threads, for the mechanical hold-down tool to snap-latch/lock onto it. The mechanical seating nipple will have female threads at the bottom, just below the latching stand-off groove. The purpose of the internal threads at the bottom end of the mechanical holding seating nipple is for using a dip tube or gas anchor

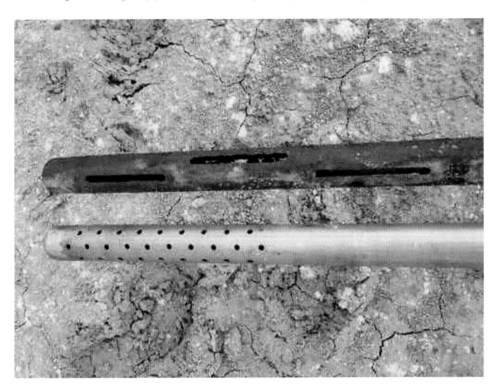

before it is screwed onto the tubing string.

It is very important to check the mechanical seating nipple to avoid running the tool upside down. Check the upright arrow (↑) on the seating nipple to avoid mistakes. I recommend that pump suppliers extend and space the mechanical tool one and a half inches longer at the bottom for better spacing purposes. (The spacing should be farther apart from the mechanical latching device at the bottom of the seating nipple to use a threaded bushing and dip tube without interference with the mechanical latching tool.)

Running and pulling the mechanical hold-down pump is similar to doing so with the standard cup-type hold-down insert pump.

It is a simple forward design (snap latch metal to metal seal landing and straight pulling action on an upward lift to unseat the pump). The mechanical pump is effective and efficient in deeper wells. It is normally run in mid-range or deeper wells because of the hydrostatic head of the column of fluid (deeper than 7,000'). The hydrostatic head is the high column of well fluid that exerts pressure on the pump at the bottom (may be several thousand pounds in deep wells).

$$HP = .0517 \times \text{pump depth} \times \text{fluid density}$$

On the mechanical hold-down insert pump, the barrel extends above the seating nipple assembly, similar to the standard cup-type hold-down pump (hydrostatic pressure on the pump barrels is less above the seating nipple). Clean wellbore fluid is recommended for all the insert pumps to avoid getting the pump stuck inside the tubing string. Using a pack-off below the pump discharge valve is recommended to prevent solids from falling around the pump barrel. Using a shear sub and/or on–off tool in deeper wells with solids is recommended.

Distinct Advantages of the Mechanical Hold-Down Pump

- Easy to latch in and seat the pump (snap latch) (see latch-on procedure)
- Easy to pull and unlatch from mechanical seating nipple (straight pickup)
- No gas anchor or long dip tub directly attached to the mechanical hold-down pump
- Gas anchor

can be screwed below the mechanical hold-down seating nipple

on the tubing string using a threaded bushing
- No seating cups to lose in the well because of the hydrostatic head of the fluid

Standard insert pumps may be designed based on the wellbore conditions.

- The bottom hold-down insert pump (RWBC)
- The top hold-down insert pump (RWAC)

On the standard bottom hold-down insert pump, the hold-down cups are spaced and made up on the standing valve at the bottom of the insert pump. The bottom hold-down insert pumps can be designed based on the wellbore conditions as follows:

- Stationary barrel with traveling plunger
- Traveling barrel with stationary plunger

Bottom hold-down pumps perform well in low-fluid and/or high-fluid wells. Artificial fluid beam pumping can reduce the bottom hole pressure to zero. The bottom hold-down insert pump is not recommended for sandy wells (will become stuck). The performance of this pump in deep wells with several thousand pounds of pressure is remarkable (without considering sand and solids). A bottom hold-down stationary barrel should be used for deep well pumping.

Figure I. Bottom hold-down pump and components.

 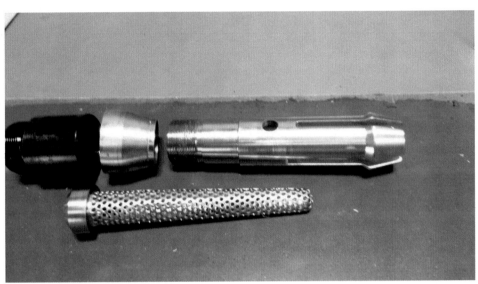

1- Valve rod bushing
2- Valve rod/pull rod (hollow or solid rod)
3- Top plunger cage
4- Plunger
5- Pin plunger
6- Ball and seat (valve cage)
7- Plug seat
8- Top rod guide
9- Top barrel bushing
10- Pump barrel (steel or bronze)
11- Cage
12- Ball and seat (travel valve)
13- Bushing
14- Seating mandrel (no-go)
15- Seating cups
16- Spacer rings
17- Seating cup nuts

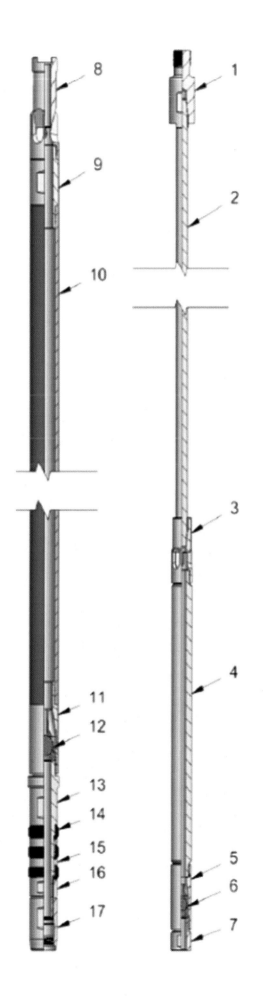

What Is the Pull Rod (Valve Rod)?

The valve rod may be called the pull rod (pulling the plunger with fluid). It is a solid rod or a hollow brass tube that connects the pump plunger to the sucker rods. Valve rods are designed with either solid carbon steel rods or hollow brass/bronze tubes.

The hollow valve

rod is made of a soft and thin wall brass or bronze hollow tube that is used as a pull rod (called the valve rod). It should be manufactured as a seamless, heavy wall tube to resist collapse. The hollow valve rod is recommended for wells with minor solids and low gassy wellbore conditions. It is also recommended to be used in travel barrel pumps

applied in shallow wellbores only. Hollow valve rod tubes may collapse because of pump-off conditions, differential pressure, and/or dogleg friction. The valves of hollow tubes are subject to fail or wear more quickly than solid rod valves.

Solid steel valve rods

are strong and tough but are subject to corrosion and friction wear. Solid pull rods must be centralized above the pump to avoid rubbing on one side.

Flat worn rod guides are due to crooked holes and rod buckling just above the pump.

(Rod buckling and friction wear can be prevented using steel and/or plastic guides to reduce pump failure.) Make sure the length of the valve rod is compatible with the pumping unit stroke length. Collapsed hollow valve rods

are mostly due to friction wear or no fluid entry to the pump chamber.

Applications of Gas Anchors in Artificial Beam Pumping

- Prevent trash from getting into the pump
- Slow down wellbore gas from entering the pump (it is difficult to avoid gas lock)

The gas anchor or dip tube is made from small carbon steel or stainless tubing to prevent trash and formation gas from entering the pump (1", 1 ⅛", 1 ¼", 1 ½"). The dip tube can be run on standard insert pumps and/or tubing pumps.

The dip tube is made up and extends below the pump, going through the seating nipple

and extending below the perforated sub (fluid intake sub)

and in the tubing string. A twenty- or twenty-four-foot gas anchor or dip tube below the standing valve is preferred in gassy wells. The gas anchor must extend to land or be located twenty feet or longer below the perforated sub on the tubing string to reduce gas interference. Using and eight- or twelve-foot gas anchor may not be a practical solution in a gassy well to avoid gas lock conditions. The application of any short strainer in low gas wells is a good practice to avoid trash from getting into the pump.

Gas anchors or dip tubes are made of either carbon steel tubes and/or steel tubes. Gas anchors are designed in various sizes and lengths and can deliver considerable volumes of fluid. A 1″ OD gas anchor may consist of 76 round holes, with 0.1875″ per hole (14.25″ opening). A 1 ¼″ OD gas anchor may consist of 68 holes, with 0.25″ per hole (17″ opening). High gas/liquid ratio wells are difficult to pump without long gas separation.

The bottom hold-down pumps may be assembled with heavy-walled or thin-walled barrels, either steel or brass/bronze, to perform in shallow and/or deep wells (800′ to 12,000′). The barrel is protected by the tubing string from external wellbore pressure. Note that all bottom hold-down pumps are designed to lift reasonably clean fluid. Pumps are not designed to suck up mud, sand, and formation solids out of a well.

Do not use the fluid pump to displace 15% or higher acid concentrations (swab the well first). The misapplication of the pumps will cost you (clean up your wellbore to cut costs). The pump barrel and plunger may become stuck or sand-cut because of improper application in any wellbore with sand, solids, and muddy fluid production (keep the wellbore fluid fairly clean).

The application of a screen as fluid intake is recommended (do not use slotted liner nipples). A shear sub or on–off tool is recommended for all bottom hold-down pumps that are designed for deeper wellbore applications (locate the shear sub four rod lengths above the seating nipple).

Using a specially made screen as fluid intake is recommended in all wells with dirty fluid. All the wells with nonconsolidated sand are recommended for a gravel-packed screen and liner. You may read up on a gravel packing procedure by this author on Google or Amazon (you will like the facts of the published material).

Pack-off rubber elements below the pump discharge are recommended to minimize solid bridging from the top and around the pump barrel and may prevent fishing work. Oil wells contain numerous problems that must be kept under consideration before installing the pump and rods in a hostile environment.

Top hold-down insert pumps

- Plunger travel (the valve rod with the plunger is directly connected to the sucker rods)
- Top hold-down pump (the hold-down cups are located on top of the pump)
- Thin-walled barrel (the wall thickness is lower)
- Cup type (it may be designed to run on nylon or composition cups)

On top hold-down pumps, the hold-down cups are made up at the top of the insert fluid pump, just below the fluid discharge valve. The pump barrel will pass through and extend far below the seating nipple and the perforated intake fluid sub. The top hold-down pump may be used as a gas anchor with or without an added gas anchor tube extending below the pump assembly (use a long mud anchor tubing joint below the seating nipple if necessary to accommodate for a longer gas anchor).

Top hold-down pumps may be recommended for traces of sand and dirty wellbore fluid. They are recommended for shallow, low-fluid-level, gassy, and emulsion oil-producing wells that need long stroke pumping. The extended barrel below the seating nipple is an advantage in semi-crooked wellbores. The top hold-down pump is preferred in low-fluid wells because of a submerged standing valve located deeper in the fluid. The pump is not recommended for deep pumping wells because of the differential pressure against the pump barrel.

Top Hold-Down Pump Components:

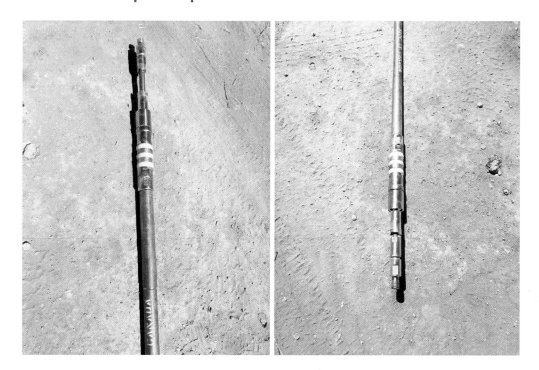

1- Valve rod bushing
2- Valve rod (pull rod, either hollow or solid)
3- Top plunger cage
4- Plunger
5- Pin plunger
6- Ball and seat (valve cage)
7- Plug seat
8- Top rod guide
9- Top barrel bushing
10- Pump barrel
11- Cage
12- Ball and seat (travel valve)
13- Bushing
14- Seating mandrel (no-go)
15- Seating cups with spacing rings
16- Seating cup locking nut

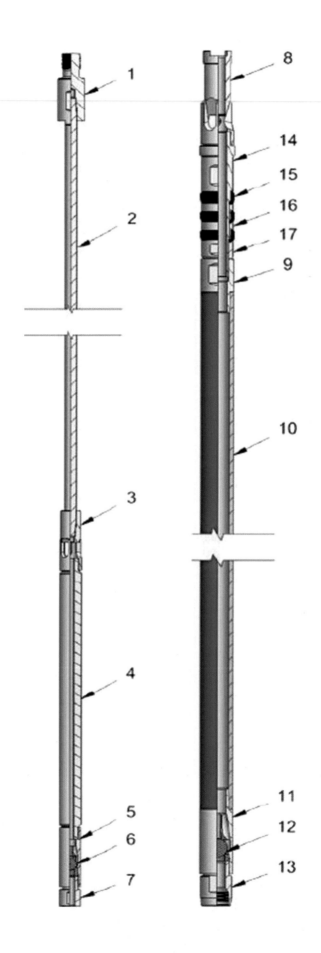

Advantages of Top Hold-Down Pumps

The fluid discharge is located directly above the hold-down cups, where the discharge fluid is turbulent. The constant movement of turbulent fluid (oil, water, and gas) that passes through the top of the pump may prevent sand and solids from bridging off on top of the pump. The extended length of the top hold-down pump below the perforated sub will act as a gas anchor.

Top hold-down pumps are useful in shallow wells that produce moderate amounts of sand, mud, silt, iron sulfide, and scale. The pump barrel may become stuck inside of the mud anchor joint

in heavy sand- and solid-producing wellbores. Top hold-down pumps are used for oil wells that are not deeper than four thousand feet because of exerted outside wellbore differential pressure and outside forces acting on the pump barrel. They are also preferred in crooked wellbores

Differential pressure is referred to as the difference between two existing pressure points in the wellbore. On top hold-down pumps, fluid cuts and/or corrosion pits outside of the pump barrel directly across/opposite of the perforated sub may be expected (because of constant fluid blasting forces coming out through the perforated sub against the pump barrel).

The "Stationary Barrel" with a Traveling Plunger Pump (RWBC)

This type of pump is the standard conventional insert pump wherein the pump barrel stays stationary, connected to the standing valve and seating cups. This pump is used for deep as well as shallow wells. The pump plunger with the traveling valve is the only part of the pump that reciprocates the rods (can be designed as either bottom hold-down and/or top hold-down pump).

The Traveling Barrel/Stationary Plunger Pumps (RWTC)

The "traveling barrel" (RWTC) is recommended for medium sand problems because the pump barrel is in constant up and down motions. This will cause fluid agitation while keeping the solids in motion. On the traveling barrel pump, the standing valve is screwed onto the plunger tube and seated stationary into the seating nipple (bottom hold-down pump). The outer barrel tube is the only part that is connected to the sucker rod string during reciprocation (the traveling barrel pump is a thin-walled barrel).

This type of pump may be used for moderate sand or solids and intermittent fluid pumping. It is not recommended the wells with gassy fluids, low fluid levels, and/or deep pumping wells (the RWTC pump works well in shallow oil wells).

RWTC (Rod, Thin Wall, Travel Barrel, Cup Type, Bottom Hold-Down)

1- Top plunger cage
2- Valve cage (ball and seat)
3- Plunger
4- Coupling
5- Pull tube
6- Coupling
7- Cage (open top)
8- Valve (ball and seat)
9- Connector
10- Pump barrel
11- Plug
12- Seating mandrel cup (no-go)
13- Seating cups (three)
14- Spacer
15- Seating cup nut
16- Bottom anchor couplings

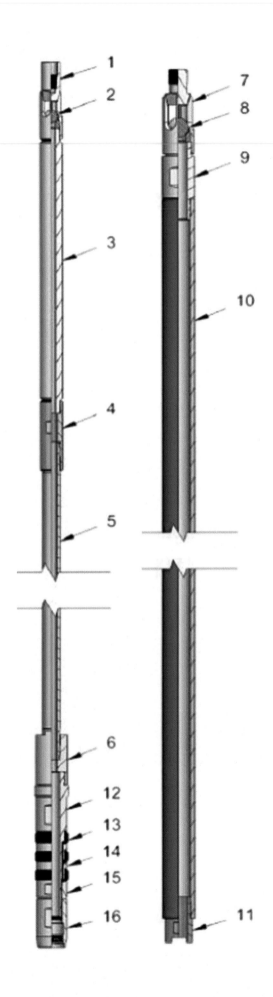

Big Bore Tubing Pumps

Artificial Lift using Tubing Pumps (THBC)

The name "tubing pump" is derived from the fact that the pump barrel is screwed and run on the tubing string (the pump barrel and the seating nipple become part of tubing string). Note that the option of using a standard seating nipple or a mechanical profile on tubing pumps is present. The purpose of a tubing pump is to displace a higher volume of fluid production (oil, gas, and water) out of a wellbore to enhance and increase oil production.

The Major Components of a Tubing Pump

The tubing pump consists of three separate components: the pump barrel (with the seating nipple), the standing valve, and the pump plunger assembly.

a) The heavy-walled large bore brass/bronze and/or steel working barrel and the seating shoe are run on the production tubing string (that is why it is called the tubing pump). The seating nipple or the seating shoe is run at the bottom of the working barrel and may screwed onto the mud anchor tubing joints below the tubing pump.

b) The standing valve and the pump plunger are separate pieces by themselves. The standing valve may be run in three different ways:

> You may drive and seat the standing valve into the seating nipple at the pump shop before the barrel with the seating nipple is run into the well.

> You may loosely screw the standing valve at the end of the pump plunger and run it in the hole on the sucker rods.

> You may choose to drop the standing valve by itself through the tubing string full of fluid before running the rods with the plunger.
> The choice of how you may run the standing valve is yours.
> You may learn the advantages and disadvantages of each method

   c) The pump's plunger is a separate piece by itself as well; it is run and travels on the sucker rod string with a small threaded retrieving screw tap (peanut thread).

Tubing pumps are designed to displace or produce more fluid than the standard insert pumps. They are also designed to operate in shallow to medium wellbore depths of six thousand feet because of the heavy fluid load (the dynamic load) on the sucker rods and the surface beam gear reducer unit. Tubing pumps are not recommended for sandy or gassy wells. I will describe the running and pulling of tubing pumps soon.

The Mechanical Principle of Subsurface Fluid Pumps

How a Fluid Pump Works

At the bottom of the insert pump barrel is a standing valve, which is dressed off with nylon or composite seating cups from the outside (the seating cups are designed to hold the pump in position and prevent the fluid from leaking). The standing valve is loaded with single or double steel cages with balls and seats internally (the balls and seats are referred to as valves).

Inside of the insert pump barrel, just above the standing valve, is a heavy-walled hollow pump plunger, which is loaded with steel alloy cages of single and/or double balls and seats, referred to as traveling valves (the cages of traveling valves are screwed at the bottom of the pump plunger assembly). The plunger assembly is screwed at the bottom of a pull rod (referred to as the valve rod), which extends from the plunger and connects onto the sucker rods just above the fluid pump assembly (the pull rod/valve rod can be a solid steel rod and/or hollow brass/bronze tube).

The standing valve and the pump plunger are spaced closely inside the pump barrel but are not connected to each other (the inset pump is made similar to a closed assembly; you cannot see the parts from the outside of the tube by looking at it). On the insert pump, the entire fluid pump assembly is connected at the bottom of the sucker rods, run through the tubing string, and seated stationary inside a polished seating nipple on the production tubing string.

As the pumping unit moves up and down, the connected sucker rods and pump plunger will stroke/reciprocate, causing the valves (standing valve and traveling valve) to operate and displace oil, gas, and water on the upward motion to the surface. An important note:

the pull rod (valve rod) on the pump plunger must be longer than the pumping unit stroke and strong enough to accommodate the dynamic load above.

Valve rods are either hollow

brass/bronze tubes and/or plain solid carbon steel rods of various sizes. Solid valve rods are used in the insert pumps of stationary barrels applied in a well at any depth range. The hollow pull tube is normally used in traveling barrel pumps in shallow pumping wells.

Hollow pull rod tubes are good for sandy and dirty fluid in corrosive fluids. The surface pumping unit and the subsurface fluid pump each have their own stroke length, which must be compatible. Note the length of the valve rod in a subsurface fluid pump.

If the pull rod on the subsurface fluid pump is shorter than the pumping unit stroke length, it will cause the fluid pump (standing valve) to pull out of the seating nipple on the up stroke (the fluid will fall back). The action of seating and unseating will damage the standing valve cups, and it will be difficult to respace the rods. The pump may lift some fluid on the up stroke.

You may have to change the stroke length on the pumping unit or pull the fluid pump out of the well to correct the error on the valve rod length. Always check the subsurface pump stroke and compare it to the pumping unit stroke length before you build the pump. The subsurface pump stroke must be longer than the pumping unit stroke.

Reservoir Pressure versus Fluid Hydrostatic Head

(Read the contents carefully.)

The reservoir pressure will force water, oil, and gas through open perforations into the wellbore. Oil and water will continue to rise up the casing and the production tubing string and may stop at a certain height. The static head of oil and water up the tubing and annulus above the perforated reservoir may reach equilibrium with the hydrostatic fluid head.

$$HP = .052 \times density \times height$$

The well fluid may rise and level off at certain height in the well equal to the reservoir pressure. If the reservoir pressure is higher than the column of fluid in the wellbore, then the well fluid will continue to flow to surface (up the casing and tubing string). In this case, the reservoir pressure is greater than the wellbore's hydrostatic fluid head.

The height of the static column of fluid depends on the reservoir bottom hole pressure. The initial height of the static fluid level will reach the same in the tubing string and annulus. The hydrostatic pressure in deep wells may reach 5,000 psi or higher. In some wells, the fluid may constantly flow without high pressure buildup out of the casing and tubing annulus because of an abnormal wellbore's mechanical problems, such as a hole and/or holes in the casing, or caused by water flood injection wells (this is an artificial reservoir flowing pressure).

The need to squeeze off casing leaks is to prevent iron sulfide bacteria and to reduce corrosion and the risk of parted tubing.

Water injection or water flooding in oil fields is the most damaging technique of increasing production without proper treatments. Why? You are basically introducing and spreading bacteria throughout the field, causing severe corrosion and parted rods. The constant flow of black iron sulfide out of the well's fluid is the major cause of pump failure. High rates of and high-pressure water injection will sweep formation sand, mud, and black bacteria into the wellbores, causing unmanageable problems.

➢ If the reservoir pressure is greater than the hydrostatic fluid in the tubing and casing string, the well will continue to flow naturally.
➢ If the hydrostatic column of fluid in the wellbore is greater than the reservoir pressure, the fluid will stall and must artificially be lifted up the hole.

THC (Tubing, Heavy Wall, Cup Type, Hold-Down)

1- Top open cage (fluid outlet)
2- Valve (ball and seat)
3- Connector
4- Plunger assembly
5- Cage (closed)
6- Ball and seat cage
7- Puller valve with spring (peanut)
8- Tubing coupling (collar)
9- Extension
10- Barrel coupling
11- Working barrel (steel or brass/bronze)

12- Extension

13- Seating nipple

Note: You may have an option to use mechanical hold-down on the tubing pump

 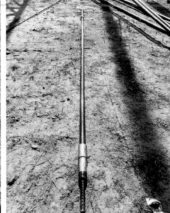 

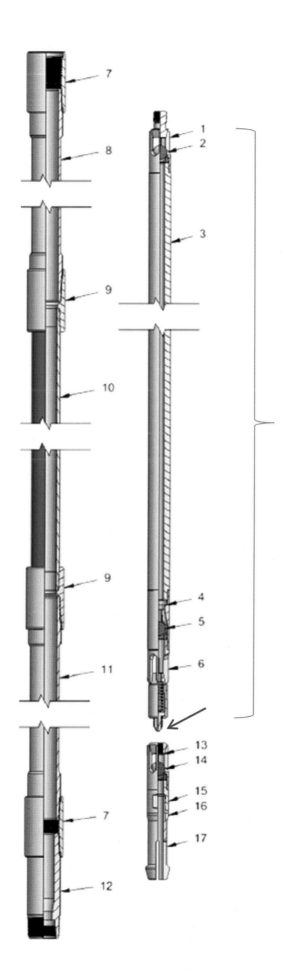

Subsurface Fluid Pump

Understanding the Functions of Standing and Traveling Valves

Subsurface valves (ball and seat) are the heart of any fluid pump. The fluid pump consists of a hollow polished plunger and a cylindrical barrel with two distinct valves (ball and seat–type valves):

- The traveling valve
- The standing valve

The traveling valve is the lifting or discharge valve, which is attached to the pump plunger or traveling pump barrel. Its function is to retain the fluid in the tubing string on the upstroke motion. On the upstroke motion, the traveling valve will close shut, lifting the fluid in the tubing string up the hole, while at same time and motion, the standing valve opens, allowing the well fluid to enter from the well or annulus and fill up the pump chamber with oil, gas, and water (creating suction/a vacuum).

The traveling valve can be viewed as the fluid discharge valve (similar to artificial swabbing cups to retain fluid). On the down stroke, the traveling valve will be opened, allowing the fluid to pass through the valves, while the standing valve is closed shut, preventing the fluid from falling back into the well. The standing valve is the fluid intake valve similar to an acting check valve. This continuous motion of stroke and reciprocation will cause the valves to close and open, displacing oil, water, and gas on the upward motion to the surface. The standing valve is a stationary valve (check valve), while the traveling valve acts as the discharge valve, traveling with the pump plunger assembly up and down with the reciprocating motion of the sucker rods.

Valve actions can be viewed on dynamometer cards. Standing valve and traveling valve motions are numerous during pumping cycles given differential pressures, fluid intake, and many other downhole conditions. When you run a dyna card, you may get one of the three hundred pumping cycle cards that need your interpretation to improve pump efficiency or find out why the pump is not lifting fluid (rod parting, valve leaks, or holes in the tubing).

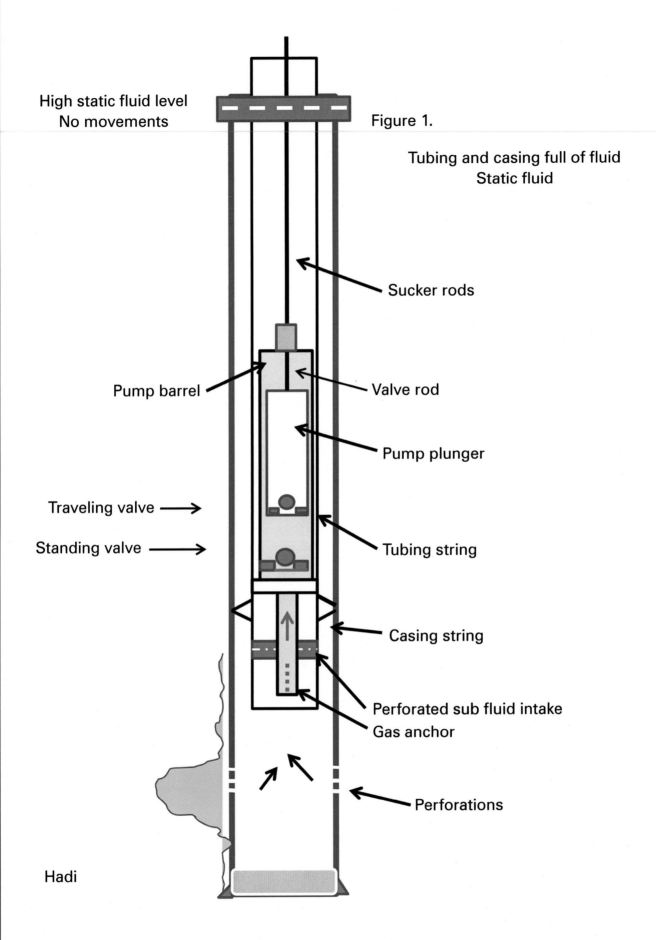

High static fluid level
No movements

Figure 1.

Tubing and casing full of fluid
Static fluid

Sucker rods

Pump barrel

Valve rod

Pump plunger

Traveling valve →

Standing valve →

Tubing string

Casing string

Perforated sub fluid intake

Gas anchor

Perforations

Hadi

KHOSROW M. HADIPOUR

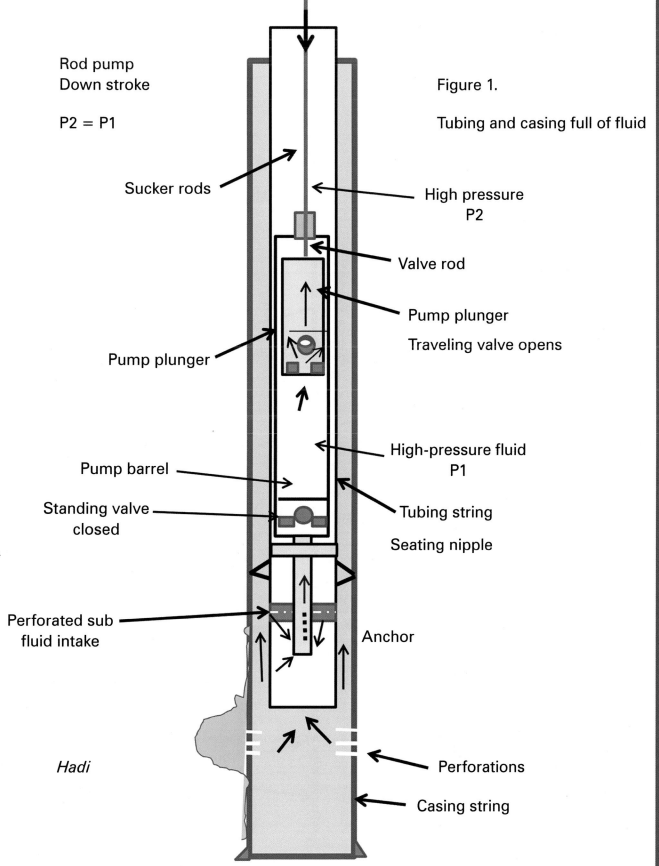

Rod pump
Down stroke

P2 = P1

Figure 1.

Tubing and casing full of fluid

Sucker rods

High pressure
P2

Valve rod

Pump plunger

Traveling valve opens

Pump plunger

High-pressure fluid
P1

Pump barrel

Tubing string

Standing valve
closed

Seating nipple

Perforated sub
fluid intake

Anchor

Hadi

Perforations

Casing string

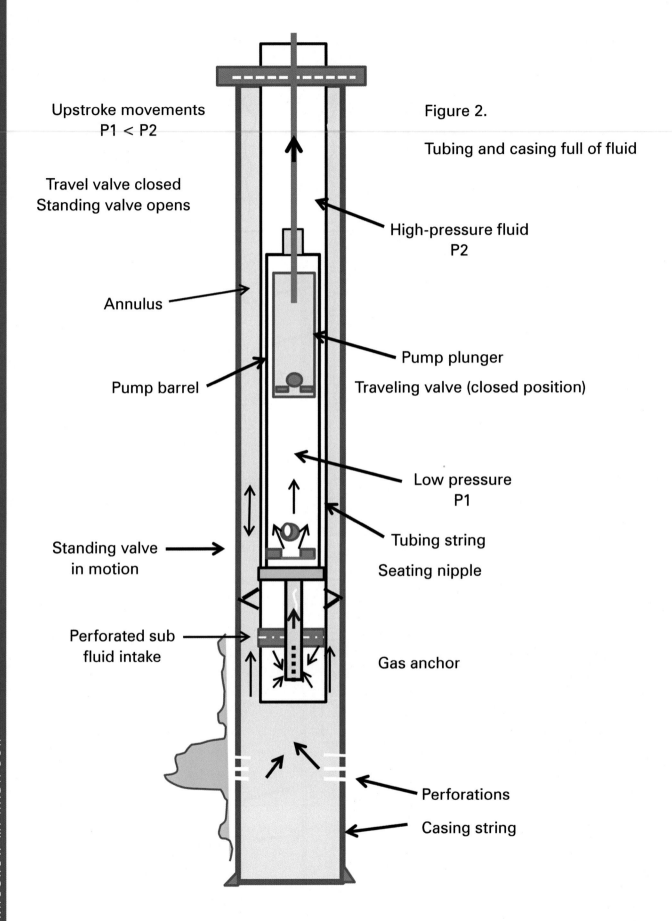

Upstroke movements
P1 < P2

Travel valve closed
Standing valve opens

Figure 2.

Tubing and casing full of fluid

High-pressure fluid
P2

Annulus

Pump plunger

Pump barrel

Traveling valve (closed position)

Low pressure
P1

Standing valve
in motion

Tubing string

Seating nipple

Perforated sub
fluid intake

Gas anchor

Perforations

Casing string

The Echo Meters

The Dyna Cards

Echo meters are very useful tools in oil and gas operations.

Echo meters are used to analyze fluid level in tubing and/or annulus and to optimize oil and gas production

The dynamometer may be a useful tool to evaluate downhole pumping cycles and measure loads. There are basically six principal loads to measure during beam pumping:

a) Peak polished rod load (referred to as maximum load)
b) Minimum polished rod load
c) Standing valve load (if any)
d) Traveling valve load
e) Counterbalance effect
f) Reference line (referred to as "the -0- line")

Using a dynamometer may enable you to diagnose and record some of the daily major and minor changes that take place during the pumping reciprocating cycling.

Dynamometers are used to collect basic occurring loads with time. Dynamometer tools may seldom repair and/or correct any downhole problems without remedial workover operations (they just indicate the sequence of event or changes). The interpretation of the cards is often questionable and subject to reviews.

Dynamometer card operation is sensitive to noise and may record over three hundred different shaped cards. Operating a dynamometer is just similar to operating an international telephone switch board or party line with too many sounds and signals coming into the system. New representative dynamometer cards should be run immediately after installing a new pump and equipment in a wellbore to obtain a fresh ideal reference card. Future dyna cards should be run after to compare any major changes.

Parted rods, holes in tubing, stuck pumps, pump-off, gas lock, and counterbalance comprise some information that may be obtained from dyna card readings. When a well is not lifting, you have to pull the well and correct the downhole problem regardless of how many dynamometer cards you may or may not run. All the surface and subsurface components on beam pumping

operations must function properly and relatively problem free to artificially lift well fluid efficiently.

Sucker rod beam pumping has a good range of pumping performance, ranging from seven hundred to twelve thousand feet deep and capable of displacing as little as one barrel to one thousand barrels of liquid per day (based on the downhole conditions and the reservoir fluid deliverability).The deeper the pumping depth, the less the beam pump efficiency.

Productivity of an Oil and Gas Well

The productivity of any artificial fluid lifting method depends on several factors:

- ✓ Reservoir deliverability characteristics
- ✓ Artificial lift design
- ✓ Artificial lifting method
- ✓ Pumping and/or lifting depth

On a beam pumping unit, there are many moving parts that must function properly to efficiently displace fluid. Several factors will work against effective and efficient beam pumping:

- Gas interference (**serious problem**)
- Low fluid (**fluid pound and water hammering**)
- Sand and solids (**plunger stick and parted rods**)
- Bad design (incorrect rod designs)
- Over travel

- Under travel

Any of the above listed items are big factors that reduce productivity. When you are standing in front of a pumping unit in motion, you may detect nearly twelve types of information without running a dynamometer. You may be able to detect surface and downhole information simply by looking and listening and filling the polished rod movements (if you are knowledgeable enough).

a) The weight falls too fast on the down stroke and very slow on the up stroke.
b) The polished rod is sticky coming up but will not fall fast enough.
c) The polished rod is very hot.
d) The polished rod is cool.
e) There is burned packing rubber on top of the stuffing box.
f) At the valve bleeder, the well is sucking on the up stroke (**possible traveling valve leak or parted sucker rods**).
g) At the valve bleeder, the well blows on the up stroke and is sucking on the down stroke (**the standing valve is leaking or the fluid pump gas-locking**).
h) At the valve bleeder, you observe suction on the up stroke and suction on the down stroke (**possible hole in the tubing or parted rods deep in the well**).
i) At the valve bleeder well blow on up stroke and blow on down stroke (**well tend to flow**)
j) If the well flows with a high gas/liquid ratio, the standing valve and the traveling valve will not function, and they will stay open (valve **chattering acting as dummy valves**).
k) If the rod tapping at the bottom (incorrect spacing, gas locking, fluid pound or gas lock).
l) If the rods tapping at the top (incorrect spacing or possible obstruction).

m) If the rods is not falling it is an indication of:

- Parted rods at shallow depth
- Stuck pump or sucker rods
- Parted tubing (crooked tubing)

Some companies may use dynamometers to diagnose pump action, and some operators will use some practical techniques.

What Is Over Travel and Under Travel?

Over travel is the condition in which the traveling valve tends to keep going further in the same direction after the sucker rods

have changed to the reverse direction.

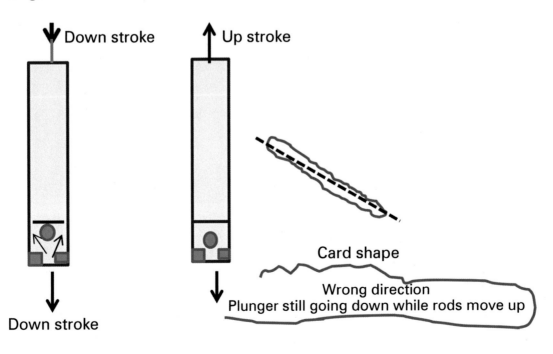

Down stroke | Up stroke

Down stroke

Card shape

Wrong direction
Plunger still going down while rods move up

Over travel may be explained as a condition that is caused by the acceleration of sucker rods in the down stroke that forces the pump plunger to travel further. Nonsynchronous pumping speed, the harmonic vibration of the rods, crooked holes, and/or downhole friction may be some of the cause and effect factors.

Characteristics of Rod Pump while in Operation

- Keep the wellbore as clean as possible. Increase pump efficiency with clean fluid.
- The sucker rod pump will have the same pump capacity regardless of pumping depth.
- If the volumetric capacity of the fluid pump is at 80% or higher, then the pump efficiency is considered to be good (if it is lifting or working well, leave it alone).
- The volumetric efficiency will increase when the annulus gas pressure is bled off or vented (reduce the annulus gas if possible).
- When free gas is allowed into the pump intake, the total volume of production will be reduced (the pump chamber will be occupied by gas).
- The standing fluid level above the pump depends on the differential pressure. The pump intake must be submerged into the fluid for the pump displace fluid and avoid water hammering (keep the fluid higher above the pump).
- If the pump displaces the wellbore fluid faster than the formation can enter the wellbore, then the pump will cause fluid pounding (need to slow down or replace the pump).
- Deep submergence of the pump in some wellbores may allow sand and solids into the pump chamber and around the plunger (a stuck tubing anchor is possible).
- Maintaining the fluid level above the pump is based on pump size and strokes per minute.
- Some wells may surge (head) well fluid because of a high gas/oil ratio and/or incorrect bottom hole design configuration (the gas surge will unload and empty the tubing above the pump).
- Heading is the cycle of periodic fluid production. Heading is controlled by gas expansion that enters the pump and exits into the tubing string.
- The application of a surface back pressure valve and adjustable choke is recommended. This is subject to cleaning because of high paraffin

and formation buildup.

- Holding a back pressure of 50# to 150# on the tubing string may keep the gas in the solution and reduce heading cycles (maintain good-order stuffing box pacing).
- The narrow space between the ID of the mud anchor and the OD of the gas anchor may cause the gas in the solution to break out, causing more gas to be released into the gas anchor and up to the pump.

The productivity of a well depends on four distinct factors of well productivity:

- Reservoir deliverability
- Wellbore deliverability
- Artificial lift design performance
- Surface and subsurface equipment performance

Reservoir deliverability may be improved by the following:

- ✓ Appropriate completion methods
- ✓ Stimulating and cleaning the wellbore
- ✓ Fracturing
- ✓ Skin damage removal techniques
- ✓ Secondary recovery methods

Wellbore and surface deliverability may be improved by doing the following:

- ✓ Maintain wellbore integrity and downhole repairs.
- ✓ Clean wellbore and wellbore fluid.

- ✓ Improve the tubing string size and pumping depth.
- ✓ Use reliable and appropriate artificial lift methods.
- ✓ Remove surface restrictions through lines and separators.
- ✓ Install back pressure valves and keep them clean.
- ✓ Reduce annulus pressure as low as possible.
- ✓ Apply hot oil treatments to remove paraffin restrictions.

The above subjects must be considered to improve the PI. Proper planning and knowledge are key to reduce lifting cost and increasing productivity. The deeper the pumping depth, the less the fluid displacement per day on beam pumping because of fluid travel time per stroke, low fluid level, and many other restriction factors. The deeper the pumping depth, the less efficient the beam pumping lift. The higher the fluid level in the annulus, the more efficient the beam pumping lift. The cleaner the fluid in the wellbore, the more efficient the artificial equipment performance.

Slow strokes per minute and long strokes will extend the life of the equipment and will reduce the cost of well repairs. Artificial beam pumping can reduce the bottom hole pressure to zero. Fluid displacement does not produce continuous lift in beam pump methods (many factors may reduce pump efficiency, including pump erosion, fluid slippage, gas lock, and solids).

The characteristics of fluid movement and intermittent fluid displacement between the pumping strokes will make the pump performance less efficient. Long and slow strokes per minute on any artificial beam pumping method is preferred. For higher fluid production volumes, artificial alternatives should be implemented. Do not beat the horse to death. Slow down the unit to avoid breakdowns. Often evaluate the lifting system; change the lifting system when changing the fluid.

Components of Insert Pump

[Use picture with all the components.]

The insert pump is made of a piece of a long hollow tube, referred to as the pump barrel, made of a brass or bronze alloy tube and/or a steel barrel tube. Pump barrels are precision honed and chrome plated from the inside (like a mirror).

Pump Barrel Tube

Barrel tubes are manufactured as thin-walled barrels (RW and RS) and/or heavy-walled barrels (RH).

Barrel:

- H — Heavy-walled barrel for metal plunger
- W — Thin-walled barrel for metal plunger
- P — Heavy-walled barrel for soft-packed plunger
- S — Thin-walled barrel for soft-packed plunger
- X — Heavy-walled barrel for metal plunger

The application of a pump barrel is to hold and protect the plunger and valves internally in a closed system while in motion as well as to bypass and deliver the produced fluid up the tubing string. The internal space between the inside of a pump barrel and the fluid plunger may range from 0.001 to 0.006 inches. The space tolerance between the plunger and the barrel may be noted as 1, 2, 3, 4, 5, and 6. Fluid slippage, fluid viscosity, and friction factors must be considered in rod pump selection. A higher-viscosity well fluid requires larger tolerance between the plunger and the barrel.

On the insert pumps, the pump barrel assembly is screwed and run at the bottom of the sucker rod string and run through the tubing string to predetermine pumping depth. The pump will be force-seated inside of a short special heavy-walled polished nipple, referred to as the API seating nipple. The purpose of the seating nipple is to hold the fluid pump stable and stationary for the duration of the pumping operation and keep the cups seated to prevent the fluid in the tubing from leaking off (seating nipples are thicker and stronger than the tubing string).

Note: [Experimentation]

If the pump is seated in a tight or crimped tubing above the seating nipple, the pump will function and will displace the fluid as long as the fluid level in the annulus stays high above the pump (you will notice small suction on the down strokes). When the fluid is pumped down near the tight spot, the pump will not displace the fluid (the pump action will indicate a hole in the tubing string). If the ball and seat are run upside down, the pump will not displace the fluid. If the pump barrel has a hole or is split, the pump will not displace the fluid. If the pump barrel is split, the pump will not lift properly.

Section IV

The Purpose of a Seating Nipple

The seating nipple is a special machined heavy-walled steel nipple that is smooth internally to land the fluid pump and the seating cups to isolate fluid leaks.

There are two types of seating nipples (used for sucker rod pumping):

- Standard seating nipple — Machined heavy-duty seamless nipple 1.10' in length and external eight-round threads on both ends (the external threads are called male threads). The standard seating shoe is a heavy-walled nipple with internal eight-round threads (the internal threads are called female threads).

- Mechanical hold-down seating nipple — A seamless heavy-duty machined steel pipe product that is designed for a mechanical hold-down pump with internal and external threads only. The mechanical seating nipple is designed especially for mechanical hold-down pumps. The length of a mechanical seating nipple is 7 ¼" long from end to end and may need to be extended to at least 1.5" longer at the end for quality improvement and sufficient space for a bushing to install a gas anchor.

The mechanical seating nipple is shorter in length, threaded externally on both ends of the nipple, and threaded internally at the bottom only for the gas anchor (or dip tube). It has a built-in machined standoff ring near the bottom for the mechanical hold-down tool and at the bottom of the pump to latch onto (see the stamped arrow on the nipple for the correct application). It is also internally threaded at the bottom for a dip tube or gas anchor using a thread bushing. Mechanical pump seating nipples are available for bottom and/or top hold-down pumps.

Pump Barrels

The pump barrel is made of either steel or brass (thin-walled or heavy-walled). Steel barrels are designed with stainless steel, chrome-plated alloy steel, or nickel carbide–coated steel (harder material). Brass/bronze barrels are generally made of a nickel carbide alloy tube that is polished, honed, and chrome-plated internally (softer material). Pump bore sizes may range from 1.062" to 3.750" (1.062", 1.250", 1.50", 1.750", 2", 2.250", 2.50", 2.75", 3.25", and 3.750").

Fluid Pump Plungers

A pump plunger is made of a hollow thick-walled carbon steel bar that is straightened, polished, and chrome-plated with a smooth outer diameter to fit inside a pump barrel. As mentioned earlier, the tolerance between the plunger and the pump barrel is measured to be 0.002 to 0.006 (2, 3, 4, 5, 6) depending on the wellbore conditions. Larger gap and tolerance between the pump and the plunger will cause high fluid slippage.

Note that a reasonable amount of fluid slippage between the pump barrel and the plunger is necessary to lubricate the plunger during reciprocations and may assist in preventing the plunger from encountering abrasive sand and solids. Plungers are designed with hard alloy and/or Monel to resist hostile corrosion and downhole abrasion.

Fluid slippage may be calculated using the equation below:

(Slippage cubic inch per minute) $Q = \dfrac{13{,}540{,}000 \ (DPC3)}{VL}$

> D = plunger diameter, inches (2.25")
> P = differential pressure, pound per square inch (2,000)
> C = tolerance between plunger and pipe (.005)
> L = length of pump plunger (48")
> V = fluid viscosity, centipoise (4)

$Q = \dfrac{13{,}540{,}000 \times 2.25 \times 2{,}000 \times (.000000064)}{4 \times 48 = 192} = 3{,}899.52$

Q= 20.31 cu in fluid leak-off   231 cu in = 1 gallon of water

KHOSROW M. HADIPOUR

As mentioned before, there are several types of plungers. Pump plungers are manufactured in different sizes and shapes:

- ✓ Pin-end spray metal plunger
- ✓ Box-end spray metal plunger
- ✓ Box-end spray grooved plunger
- ✓ Monel metal plunger
- ✓ Ring-type plunger

  The rings on the plunger could be composition-type or split rings,

- Plain chrome-plated plunger — Made with high-quality chrome alloy with threaded endpoints.
- Nylon rings–type plunger — Packing ring–type plunger to handle sand and solids as well as resist corrosion attacks.
- Grooved plunger — Built-on grooved plunger to perform in fluid with solids.

Short-length pump plungers should be used to effectively displace viscous fluids.

Pump Plunger Diameter Size (D)	Plunger Constant (C)
¾"	0.066
⅞"	0.089
1"	0.117
1 1/16"	0.132
1 ¼"	0.182
1 ½"	0.262
1 ⅝"	0.308
1 ¾"	0.35
1 25/32"	0.371
1 ⅞"	0.440
2"	0.466
2 ⅛"	0.526
2 ¼"	0.591
2 ½"	0.728
2 ¾"	0.882
3 ⅛"	1.136
3 ¼"	1.231
3 ¾"	1.639
4 ¼"	2.106
4 ¾"	2.630

Pump Cages

The pump cage is designed to hold and support the valves (balls and seats) while lifting and falling in place. There are a variety of sizes and engineered cages for various wellbore conditions. Cages are designed with high-quality low-carbon steel, steel alloy, stainless steel, and high-quality Monel (copper/nickel alloy).

Pump cages

are designed from **nonmagnetic** and heavy-duty steel material. Traveling valve cages are made as closed-type or open-type cages. Standing valve cages are made as closed-type cages for the insert pumps

and standing-type cages for the tubing pumps.

Balls and Seats (Valves)

Valves are critical parts of any artificial fluid pump. The success or failure of any downhole pump is dependent on the selection and quality of balls and seats that are put in a pump (valves are actually **the heart** of a fluid pump). Never cut costs on the balls and seats. Put the best ball and seat

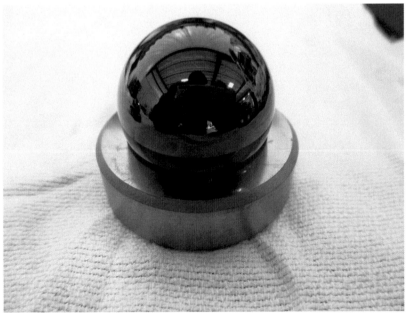

in the pump based on the wellbore conditions (a local pump shop should know that).

The balls and seats are manufactured from various high-quality and fine-polished materials to meet or exceed the API requirements. Rigid, precision machining, polishing, and vacuum testing are very important in ball and seat manufacturing techniques (local experience of pump manufacturing). Carry a vacuum gauge to test your pump before running it in any well.

Stainless Steel Ball and Seat Configurations

The stainless steel seat size ranges from 1 1/16″ to 3 ¾″ in diameter (1 1/16″, 1 ¼″, 1 ½″, 1 ¾″, 2″, 2 ¼″, 2 ¾″, 3 ¼″, and 3 ¾″). The ball size may range as follows: ⅝″, ¾″, 15/16″, 1 ⅛″, 1 ¼″, 1 11/16″, 2 ⅛″, and 2 ¼″). There are alternate balls that may be used in special made-up pumps.

Ball and Seat Actions during Fluid Lift

Seats are designed with flat, polished, and chromed surfaces, slightly beveled for the ball to land and seat without any gaps (complete seal-off configuration). The valve sits normally at rest without significant up and down or rotational movements. Light balls are normally at constant lift or rotational movements during opening and closing configurations (chattering motions). Balls and seats can become damaged because of the length of pumping time, fluid movements, corrosion, fluid hammering, and solid abrasions. The misapplication of balls and seats will cause deep abrasions and fluid leaks.

Balls are manufactured to a near-perfect spherical shape, polished, and chromed. Balls constantly move during fluid lifting. Lightweight balls will have constant movements and rotation, while heavy balls may only be lifted up and down based on fluid input into the pump chamber. Balls are designed to withstand hostile wellbore environments (temperature, fluid velocity, gas, sand, and solids). Depending on fluid height and intake pressure, the ball is in constant rotational, up and down movements while opening and closing inside the cage.

Balls and seats can be intermixed for different application purposes:

- Fluid corrosion (H2S and CO2)
- Solid abrasion
- Shallow pumping depth
- Deep pumping depth
- Formation sand and iron sulfide
- Acid and chemical applications
- Fracture sand, formation solids, and gyp

Ball and Seat Selections

➢ Tungsten carbide seat and silicon nitride balls: Nitride balls are lightweight and hard with fracture toughness and strength. The combo has high resistance to corrosion and abrasion.

➢ Tungsten carbide seat and titanium carbide balls: Titanium carbide balls are lightweight, hard, and resistant to corrosion and abrasion.

➢ Nickel carbide seat and silicon nitride ball: These have excellent resistance to H2S and CO2 corrosion.

➢ Titanium carbide seat and ball: They are lightweight, with high resistance against abrasion.

➢ Silicon nitride seat and ball: These are lightweight, with high mechanical strength, and resistant to deformation.

All the balls and seats must be selected and implemented based on the wellbore conditions.

Fluid Pump Makeup and API Identifications

Learn to read pump makeup identifications before running the pump in the well:

An RXBC pump means the following:
R = Rod type
X = Heavy-walled barrel
B = Bottom hold-down
C = Cup-type bottom hold-down pump

An RTHC pump means the following:
R = Rod type
T = Tubing pump
H = Heavy-walled barrel
C = Cup-type standing valve

An RWAC pump means the following:
R = Rod type
W = Thin-walled pump barrel
A = Top hold-down pump
C = Cup-type standing valve

An RWTC pump means the following:
R = Rod type
W = Thin-walled barrel
T = Traveling barrel
C = Cup-type hold-down pump

An RHBC pump means the following:

R = Rod type
H = Heavy-walled barrel
B = Bottom hold-down pump
C = Cup type

An RHBM pump means the following:

R = Rod type
H = Heavy-walled barrel
B = Bottom hold-down pump
M = Mechanical hold-down pump

Example:

If a pump card shows "2,500 × 1,500 × 24' RXBC-3'," it means the following:

2,500 = 2 ½" or 2 ⅞" tubing
1,500 = 1 1/5" pump plunger
24' = Twenty-four-foot-long pump barrel
3' = Three-foot-long plunger
R = Rod type valve rod
X = Heavy barrel
B = Bottom hold-down pump
C = Cup-type seating elements

The pump bore (plunger size) designations are as follows:

125 = 1 ¼"
150 = 1 ½"
175 = 1 ¾"
178 = 1 25/32"
200 = 2.0"
225 = 2 ¼"
250 = 2 ½"
275 = 2 ¾"
375 = 3 ¾"

The fluid pump is either a rod pump (R) or a tubing pump (T).

The fluid pump barrel designations in terms of wall thickness are as follows:

H = Heavy wall thickness
W = Thin wall thickness

The fluid pump can be a top hold-down pump (A), a bottom hold-down pump (B), or a bottom hold-down traveling barrel (T).

The seating nipple identifications are as follows:

C = Cup-type pump
M = Mechanical hold-down pump

The tubing designations in the oil field are as follows:

15 = 1 9/10" tubing
20 = 2 ⅜" tubing
25 = 2 ⅞" tubing
30 = 3 ½" tubing

Pump Plunger Diameter Size (D)	Plunger Constant (C)
¾"	0.066
⅞"	0.089
1"	0.117
1 1/16"	0.132
1 ¼"	0.182
1 ½"	0.262
1 ⅝"	0.308
1 ¾"	0.35
1 25/32"	0.371
1 ⅞"	0.440
2"	0.466
2 ⅛"	0.526
2 ¼"	0.591
2 ½"	0.728
2 ¾"	0.882
3 ⅛"	1.136
3 ¼"	1.231
3 ¾"	1.639
4 ¼"	2.106
4 ¾"	2.630

❖

Subsurface Pump Displacement Calculations

The production calculations are based on the theoretical capacity and the actual pump capacity. The pump bore multiplied by the length of plunger travel will give you the cubic inches of fluid per stroke.

231 cubic inches = 1 gallon

<u>Example:</u>

Pump plunger 2.25″ = 2 ¼″

Strokes per minute = 8

Stroke length = 160″

<u>Pump Plunger Constants:</u>

<u>Size</u>		<u>Factor</u>
1 ¼″		.182
1 ½″		.262
1 ¾″		.357
2″	**	.468
2 ¼″		.59
2 ½″		.728
2 ¾″		.88
3 ¼″		1.231
3 ¾″		1.64

Theoretical Pump Displacement Calculation

24 hours of fluid displacement = strokes per minute × stroke length × pump constant

<u>Example I</u>

Pump stroke = 8 strokes/minute

Stroke length = 160″

Plunger size = 2″

24 hours of pump displacement = 8 × 160″ × 0.468 = 599 barrels per day at 100% efficiency

755 × 80% = 479 barrels per day (may be closer to actual daily production)

The real production is the actual fluid displacement in twenty-four hours when you gauge the oil stock tank. The pump calculation consists of strokes per minute, the stroke length, and the size of the pump plunger (less stretches and over/under travel).

<u>Example II</u>

Strokes per minute = 8

Stroke length = 160″

24 hours of pump displacement = (plunger diameter)2 × SPM × stroke length × .1166

24 hours of pump displacement = 2″ × 2″ × 8 × 160 × .1169 = 599 bbl/day (100% efficiency)

599 × 0.80 = 479 bbl/day (80% pump efficiency)

Volumetric pump efficiency = <u>Actual volume</u> × 100
$\qquad\qquad\qquad\qquad\qquad$ Theoretical volume

Consider fluid gravity (G):

G = <u>(BOPD × S. gravity of oil)</u> + <u>(BWPD × S. gravity of water)</u>
$\qquad$ Total fluid per day $\qquad\qquad$ Total fluid per day

Volumetric efficiency = <u>Actual fluid volume</u> × 100
$\qquad\qquad\qquad\qquad$ Theoretical volume

The actual fluid volume is the measured fluid at the tank battery. The theoretical volume is the calculated performance of the subsurface pump based on the plunger size, pump stroke, and number of strokes per minute.

Wellbore Preparation before Beam Pump Installation

<u>Workover Operation</u>

Pulling and Running Beam Pumping Artificial Lift Equipment

- ✓ Clear the surface location for the incoming workover rig and equipment.
- ✓ Install and test the safety anchors to twenty-three thousand pounds of tension (tag and show the test date on each of the anchors after testing). Beam-type safety anchors may be used and must be certified for safety purposes.
- ✓ All the safety anchors on the well location must be tested at least once every three years to stay in compliance.

- ✓ Hold and document safety meetings. Clearly discuss the work procedure to the rig crew.
- ✓ Move in and rig up the workover rig, tools, and equipment. Use steel mats if necessary. Safety steel mats are very useful to avoid sinking in soft areas near the wellbore
- ✓ Install and tie in the safety guy lines to the permanent safety anchors on the location as required (dead man).
- ✓ Move in and spot a clean pump and tank (the reverse unit equipment).
- ✓ Fill up the rig tank with 2% KCL water or clean brine water (kill fluid). Do not introduce fresh water to the wellbore without permission from the well owner.
- ✓ Check the well for pressure. Read and record the shut-in tubing and casing pressures.
- ✓ Spot H2S safety equipment if necessary (make sure all the H2S monitors are working).
- ✓ Bleed off the pressure from the tubing and casing strings into the rig tank/tanks.
- ✓ Circulate to kill the well with heavy brine water if necessary to keep the wellbore fluid static while working on the well (the well must be dead during the workover operation).
- ✓ The best wellbore killing procedure is by circulating gas and oil out of the wellbore. Bull heading a few barrels of kill fluid down the tubing and casing string may not be a satisfactory practice on some wellbores (the well may kick on you unexpectedly).
- ✓ If high H2S is present in the wellbore, the well must be kept under control to protect the rig crew. Never assume a wellbore without H2S gas.
- ✓ Always check the wellbore for flow (early detection and quick reaction are key to well control).
- ✓ Isolate the production flow lines at the well to avoid backflow. Remove the Christmas tree and/or wellhead equipment and install the TIW (safety) valve on the tubing string.

Tips:

- ✓ Note that the TIW safety valve is designed to hold pressure from the bottom only (test your safety valve regularly to keep yourself safe).
- ✓ Use low-profile tubing slips to avoid pulling too much tension into the tubing string to clear the wellhead flange and/or mechanical tension set packer (if necessary). (Check the range and capacity on all the tools and equipment before application.)
- ✓ Back off and remove the wellhead studs (bolts) and nuts. Check the well for flow before removing the wellhead equipment (make sure the well is completely dead).
- ✓ Remove the wellhead equipment safely.
- ✓ Install and nipple up the BOPs (use correct-sized ring gaskets to avoid leaks). I prefer hydraulic actuated BOP stacks versus manual BOP stacks. The manual BOP stack is too slow to open and to close, and often the rams will leak pressure. The manual BOPs are good for dead or depleted wells only.
- ✓ Test the BOP at 200 psi (low) and 3,000 psi (high) or as required (never shortcut on BOP testing).

- ✓ Release the production packer and/or TAC if any. Circulate the well if necessary (keep the hole full properly with kill fluid during the pipe trips).
- ✓ The tubing anchors and tension packers must be completely released to prevent them from being hung up during pipe trips.
- ✓ Do not allow well kicks when tripping the pipe out of the hole and/or going back into the well.
- ✓ Keep the hole full of fluid properly while tripping out the hole.
- ✓ See the releasing procedure for the packer and TACs (set and release procedures on the TAC and various types of packers).
- ✓ Check and monitor the tubing and casing string for H2S gas at all times. Never assume a wellbore without H2S gas (protect yourself against H2S poison gas).
- ✓ Pull, strap, and inspect the tubing string as it comes out of the hole (scan the tubing if necessary).
- ✓ If you are pulling the TAC, you may turn the TAC right for two rounds every twenty stands while tripping the tubing out of the hole. This may prevent the TAC from setting while pulling and backing off the tubing string to the left.
- ✓ Keep the hole full properly as you trip out the hole (pulling the tubing out of the fluid will reduce the hydrostatic head of the fluid in the casing string and may cause kicks).
- ✓ The wellbore fluid must be kept in static condition when working on any well at all times.
- ✓ Finish out the hole and lay down the excess downhole equipment (packer, TAC, etc.).
- ✓ Close the BOP shut as soon as the equipment is pulled out of the hole.
- ✓ Never leave the wellbore open during lunch breaks.
- ✓ Prepare to trip in the hole with the work string for the wellbore cleanup trip.

Wellbore Cleanup Preparation before Running an Insert Pump

- ✓ Review the downhole condition (perforating depth and plug-back depth).
- ✓ Space out and trip in the hole with a tri-cone rock bit and a casing scraper on good-condition tubing joints (space out between the rock bit and the casing scraper to avoid running the scraper through the open perforations).
- ✓ Rig up the tubing testers to the hydrostatic test and drift tubing string in the well. Read the chapter on the advantages and disadvantages of hydrostatic testing tools. Read also the available published book on hydrostatic and hydro testing in the oil field by this author (published by Xlibris).
- ✓ Check and make the bit cons free (do not hammer on the shanks) Trip in the hole with the correct-sized rock bit and casing scraper on the tubing string. Keep an accurate pipe tally on the tubing string and the bottom hole assembly while testing and tripping into the well.
- ✓ Run in the hole slowly with the bit and scraper assembly to avoid unexpected, sudden stops.

- ✓ If tagging on any tight spots or solids, do not rotate or drill with the casing scraper in the well (the casing scraper is not designed for drilling).
- ✓ Watch for downhole restrictions, such as tight spots, lost objects, screens, and liners.
- ✓ Drift and clear the casing and open perforation with the rock bit (do not push the scraper through tight perforations; the scraper must be spaced above the open perforations).
- ✓ Rig up to circulate the wellbore.

There are two methods of circulating a well:

- – Down the casing annulus and out of the tubing string (referred to as the "short way")
- – Pumping down the tubing string and out of the casing annulus (referred to as the "long way")

In remedial workover operations, we generally circulate a well down the casing and out of the tubing (short way). It is faster and safer. The reason for circulating down the casing and out of the tubing is to get the dirty sand and solids more quickly to the surface (less circulating time and volume). In workover operations, normally, non-viscous fluid is used to clean up wellbores.

Pumping and circulating down the tubing and out of the casing (the conventional method) may take a longer time to bring the dirty sand and solids out of the annulus. Circulating with non-viscous fluid always entails the risk of getting stuck in the casing when circulating a well the long way (down the tubing and out of the casing annulus).

- ✓ Always circulate the wellbore. Clean two volumes with produced water or as required (catch samples and report findings).
- ✓ Clean up the wellbore and leave a one-hundred-foot rat hole below the open perforations, if possible.
- ✓ Pool with the tubing slowly while filling up the hole up with a clear, clean fluid.
- ✓ Lay down the bottom hole assembly (bit and scraper assembly).
- ✓ The wellbore is now clean enough to run artificial lifting equipment.

Beam Pumping Operation and Installation Recommendations

These recommendations are based on many beam pumping successes and failures.

a) Clean up the wellbore as best as possible before running artificial lifting equipment.
b) Select larger "seamless" tubing if possible (a larger ID tubing string is needed to install the rod guides with sufficient space for the fluid to pass).
c) Select and use large mud anchor joints if possible (a minimum of two joints of large ID tubing is needed to obtain a minimum of one inch of space between the mud anchor ID and the dip tube gas anchor OD).

d) Select and run two perforated subs if needed (the perforated sub is the fluid intake to the pump). Do not use slotted subs (the slots will wash out and allow trash into the pump).

e) Select and use a tension-set TAC below the fluid pump (the TAC is designed to anchor the tubing string and should be landed below the pump).

f) Properly set the tubing anchor and hang the tubing string in tension (do not use the TAC in bad casing and/or high H2S-corrosive wellbores).

g) Clean up and pack off the wellhead properly to avoid leaks from the casing string.

h) Use a long stainless steel gas anchor/dip tube below the pump if possible.

i) Select and run a suitable fluid pump to lift the well fluid (avoid running a large pump and/or undersized fluid pump based on reservoir fluid deliverability).

j) Select and use the correct designed tapered rod string as much as possible or a smaller sucker rod string to be able to install protective rod guides to slow down tubing wear.

k) Select and use the best sucker rod string to reduce wellbore corrosion and wear.

KHOSROW M. HADIPOUR

l) Use good-quality rod guides

on the sucker rods to reduce friction wear (a minimum of three rod guides on each rod is recommended). I prefer using molded-on rod guides to keep the rod guides from sliding, causing parted rods and fishing work).

m) Use a pony rod

with a molded stabilizer guide above your fluid pump.

n) Use weight bars

or larger sucker rods as needed (avoid using large bars to prevent fluid choking effects and tubing friction wear- **do not use large bars inside tubing pumps).**

o) Space out the rods correctly (avoid bumping on upstroke and/or down stroke motions) Never bump on the standing valve

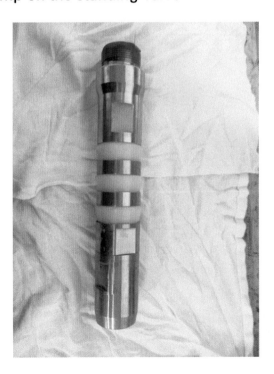

in tubing pump.

p) Counterbalance the unit properly to save energy.

q) Pack off the stuffing box properly to avoid oil and water pollution.

r) Hang the carrier bar flat and straight at the center of the horse head.

s) Slow strokes per minute with long strokes are recommended.

Here is how to run production equipment to put a well on pumping mode:

1. Rig up the tubing tester to test the tubing string in the hole if necessary (the tubing must be free of holes and tight spots).
2. Testing and drifting the tubing string is always a good practical procedure.
3. Hydrostatic testing tools will not only detect leaks but also drift the tubing string while running in the well.
4. Casing strings must be kept in good condition without holes and doglegs.
5. The long hydro testing tools are not only for checking the pipe for leaks but also for drifting tubing strings to detect tight spots and tool joint leaks as well as wiping off the tubing string internally.

Running the Tubing String on Artificial Lift Beam Pumping

1. Trip in the hole with two mud anchor joints on a blind bull plug at the bottom of the mud anchor joints. The mud anchor joints will store mud, scale, sand, and sludge and may prevent junk or trash from getting into the fluid pump's intake. All wells without gravel packing should have two mud anchor joints below the seating nipple. A wellbore without gravel packing may produce several feet of mud, sludge, and sand in the mud anchor joints (especially wells with holes in the casing). The mud anchor joints must be long enough to run a longer dip tube and/or gas anchor without reaching the bottom and cause sand and mud to enter the pump. The application of a long gas anchor is recommended in all high gas-producing wells.
2. Use a larger-diameter mud anchor joint if possible to accommodate sufficient space between the ID of the mud anchor and the OD of the dip tube for the fluid entry.
3. I do not recommend orange pill cuts below a mud anchor joint (the purpose of the orange pill pipe is to avoid using a blind bull plug and collar).
4. I do not recommend using a slotted mud anchor. The slots will wash out and are seldom of correct size (slots will allow junk into the pump). Some people cut the slots with a cutting torch, which is a poor practice and seldom works.
5. Large mud anchor joints are ideal for the gas anchor or a dip tube to pass fluid and function properly.
6. Small-diameter casing strings are major restrictions to correctly design a mud anchor (do not choke the fluid passing by the mud anchor).
7. Large mud anchor joints (tubing collars) must not be used to choke or restrict the flow from the open perforation onto the perforated subs. This will create choking effects and turbulent fluid at the tubing collar, causing fluid washouts and fluid pounding.

8. The minimum clearance between the OD of the gas anchor and the ID of the mud anchor joint must be at least one inch or larger to keep the fluid flowing into the pump with ease.
9. Two mud anchor joints (61') may serve the purpose in all low gas/oil ratio wells without any sand, mud, or solids.

<u>Example:</u>

The casing is 5 ½", and the production string is 2 ⅜" = 2.375 (1.995 ID). Thus, the mud anchor should be 2 ⅞" = 2.875" (2.259), and the dip tube should be 1.25".

2.875" − 1.25" = 1 inch clearance for the fluid to pass by the tube (½" space)

Using 3 ½" mud anchor joints in 5 ½" casing is questionable (the collar is 4.5" OD) (use a non-upset pipe if necessary).

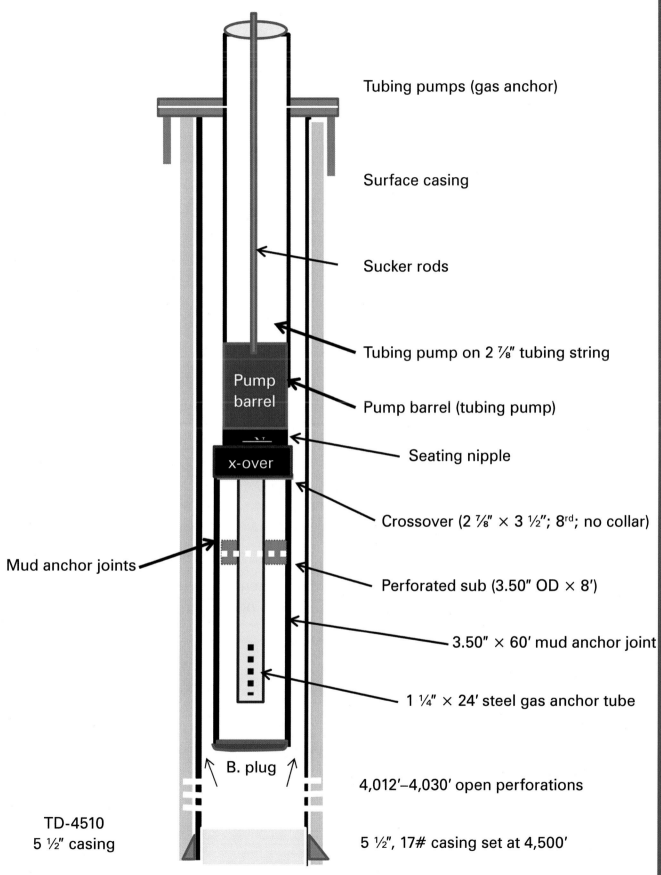

Tubing pumps (gas anchor)

Surface casing

Sucker rods

Tubing pump on 2 ⅞″ tubing string

Pump barrel (tubing pump)

Seating nipple

Crossover (2 ⅞″ × 3 ½″; 8ʳᵈ; no collar)

Mud anchor joints

Perforated sub (3.50″ OD × 8′)

3.50″ × 60′ mud anchor joint

1 ¼″ × 24′ steel gas anchor tube

B. plug

4,012′–4,030′ open perforations

TD-4510
5 ½″ casing

5 ½″, 17# casing set at 4,500′

Conditions of an Ideal Mud Anchor Joint

A mud anchor tubing joint or joints must be designed to be slick and without large tubing collars or large crossover subs to avoid fluid choking and causing turbulent fluids. The mud anchor joint must have sufficient clearance internally to run a gas anchor or dip tube with sufficient area for the fluid to flow without restrictions of any kind. It must also have sufficient clearance externally to allow the reservoir fluid to flow. It passes through the casing annulus without choking the fluid, creates holes in the mud anchor joint, and washes out in the pipe.

Running a slick large mud anchor joint/joints is ideal if it can be arranged via a safe butt-welding procedure or using tubing joints with slimehole collars. Two mud anchor joints should be used in all wells with high gas/liquid ratios to run a long gas anchor to control the gas and prevent gas locking problems. One mud anchor joint is practical in all wells with minimum gas-producing wellbores. One tubing joint may be used with orange pill cutting and welding.

There are two types of fluid pounding in a wellbore: fluid pound, often called gas pound; and water hammering (liquid pounding). It is often difficult to differentiate between the two to come up with better solutions. Gas or fluid pound is difficult to deal with within pumping wells. It often becomes serious enough that it can damage surface and subsurface equipment:

- Part the sucker rods
- Create holes in the tubing string
- Unseat the tubing anchor
- Break the polished rod

Fluid pound usually takes place in the down stroke, when the travel valve opens, releasing the fluid force on to the standing valve, causing high pounding energy. To prevent fluid pound, several changes can be made:

- Slow down the pumping unit
- Lower the pump further into the liquid phase (do not sand up the pump)
- Put the well on a time clock
- Reduce the fluid pump size
- Put the standing valve and traveling valves closer
- Improve the gas anchor and dip tube
- Reduce differential pressure

Below is the clearance between the casing and the tubing for mud anchor joints:

Casing OD	Inside Drift Diameter
4 ½" — 9.50#	3.965"
4 ½" — 10.50#	3.927"
4 ½" — 11.60#	3.895"
4 ½" — 12.60#	3.832"
4 ½" — 13.50#	3.795"

Selecting Mud Anchor Joints

2 ⅜" (2.375") OD mud anchor joint	Tubing collar 3 1/16" (.0625") (okay to use)
2 ⅞" (2.875") OD mud anchor joint	Tubing collar 3 ½" OD (choking fluid)
3 ½" (3.50") OD mud anchor joint	Tubing collar 4 ½" OD (cannot be used)

Casing OD	Inside Drift Diameter
5 ½" — 14#	4.387"
5 ½" — 15.50#	4.825"
5 ½" — 17#	4.767"
5 ½" — 20#	4.653"
5 ½" — 23#	4.545"

Mud Anchor Joints

2 ⅜" (2.375") OD mud anchor joint	Tubing collar 3.0625" OD
2 ⅞" (2.875") OD mud anchor joint	Tubing collar 3.50" OD
3 ½" (3.50") OD mud anchor joint	Tubing collar 4.50" OD (choking fluid)*

*You may use slimehole collars.

Casing Size	Inside Drift Diameter
7" — 17#	6.413"
7" — 20#	6.331"
7" — 23#	6.241"

7" — 26#	6.151"
7" — 32#	5.969"

What Size a Mud Anchor Joints Fit

2 ⅞" (2.875") OD mud anchor joint	Tubing collar 3.50" OD (okay to use)
3 ½" (3.50") OD mud anchor joint	Tubing collar 4.50" OD (okay to use)
4" (4.0") OD mud anchor joint	Tubing collar 5.0" OD (okay to use)

Note that some operators may or may not use mud anchors below the pump at all. They may only use the seating nipple below the tubing string only. (The risk of gas, solids, and trash getting into the pump is higher.)

Here is how to run the tubing landing for an insert pump (from the bottom up):

- Blind bull plug (at the bottom)
- Two joints of 2 ⅞" tubing (or as needed) Leave 20' or more space below dip tube!
- 4' or 8' perforated sub (use a 4'-long sub if there is too much gas)
- TAC double or single springs
- API seating nipple (1.10')
- The rest of the production tubing to the ground surface

Install a (4' to 8') perforated nipple on top of the mud anchor joints (fluid intake). Do not use a slotted pipe joint in the well (the slots will wash out). Standard round holes on the perforated nipple are preferred over slotted holes. Slotted holes will cut and wash out more easily than round holes. Some slotted holes are made with a cutting torch, and they are seldom accurate (allowing trash into the pump). (The perforated nipple is the fluid intake, allowing fluid to enter the tubing string and the pump.)

None on TAC Application in Beam Pumping

As mentioned in previous pages, the purpose of the TAC is to hold the tubing in tension, keep the tubing from falling into the well, and keep the tubing from moving side to side and/or wobbling. The TAC is also designed to keep the tubing from dropping in the well in case the tubing becomes parted. I prefer to install the TAC below the seating nipple, if possible.

Note that if you run a TAC above a tubing pump and/or seating nipple, make certain that the TAC is fully open for the pump and/or pump plunger to go through the TAC. The tubing anchor

will be effective at any spot below the seating nipple (below the pump).

1. Install the standard API seating nipple above the TAC.
2. Make certain that the hold-down cups fit tightly in the selected seating nipple. Using a seating shoe is an option. A seating shoe is made with internal female threads only on both ends and will save you the cost of using two tubing collars on the standard seating nipple. The standard API seating nipple for the insert pump is heavy-walled and machined smoothly internally and externally. The seating nipple is 1.10' long and made up with eight round threads externally on both ends (beveled for hold-down cups and the no-go).

3. If you plan to run a mechanical hold-down insert pump, you must use a mechanical hold-down seating nipple only (the mechanical hold-down nipple is different). You may use a gas anchor or a dip tube

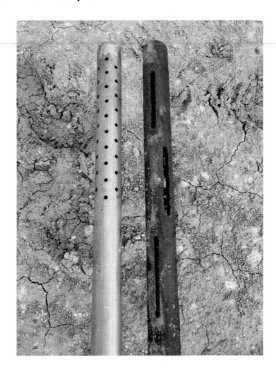

below the mechanical seating nipple on the tubing string before running the mechanical hold-down pump.

4. The mechanical hold-down seating nipple enables you to use a long gas anchor or dip tube below the mechanical seating nipple on the tubing string (if you forget to install a gas anchor below the mechanical seating nipple, you *cannot* run a gas anchor below the pump).

5. All the seating nipples must be free of pits, cuts, and rough spots internally to avoid fluid leaks. Make sure the seating nipple is API designed and not an aftermarket product. The artificial lifting equipment must be designed to fit like pieces of a puzzle.

Continue Going into the Hole with the Tubing Landing . . .

6. Apply a light pipe dope or thread compound on the clean tubing pin thread ends only. Make up a tubing string of three rounds by hand first before using hydraulic power tongs. (Use the recommended makeup torque by the pipe manufacturer.)

7. Hydro-test if necessary and run the rest of the required tubing going in the hole (report the tubing condition after hydro testing).

After a Successful Trip in the Well with Production Tubing

8. Stop and check the well for flow before removing the BOPs (running the pipe in the well will raise the fluid level higher because of tubing displacement, and it may stimulate the well, causing the wellbore gas to break out of the solution and cause well kicks).
9. Nipple down and remove the BOPs (check your well for flow before removing the BOPs).
10. Install the wellhead flange or slip-type wellhead equipment (clean up the wellhead groove and the ring gasket).
11. Do not install any tubing hanger (donut) if you are using any tension equipment such as TACs or tension packers. A tubing string without a TAC is subject to piston movements and ballooning effects.
12. Measure and set the TAC at the required depth (report setting the depth).
13. Space out, pull, and leave 5,000–10,000 lbs. of tension on the tubing anchor only (tension against anchors is based on well depth, casing size, and casing condition).

To Set the TAC . . .

Read and understand the procedure of setting and releasing TACs to stay safe and avoid damage to the well. Space out the tubing as required and lower the tubing to the measured depth below the wellhead (8" to 9" or as required). Position the tool at the setting depth and rotate the tubing five to eight rounds to the left (counterclockwise to set the tool).

Pull upward with a tension of 5,000–9,000 lbs. onto the tubing string (do not over-pull onto the tubing string). To release the TAC, slack off the weight and apply five to eight right-hand rotations to release. Some TACs are designed to set and/or to release a quarter turn at the tool using mechanical J-slot TACs. Note that on normal tubing landing operations, the standard TAC will be set to the left and released to the right (do not run the anchor upside down).

On PCP installation, the TAC will be set to the right and released to the left only. Note that H2S poison gas will penetrate and damage the bow spring on the TACs. You may lose the bow springs in the hole (it may be difficult to release the anchor). Most H2S wells are full of broken steel bow springs that are not reported. (Some wellbores with H2S are full of junk to clean up.) The application of a tubing string in beam pumping is subject to the piston effect.

Note that on PCP downhole pump installations, the TAC may be set to the right and released to the left (I will explain the reason later). (Read PCP landing reports carefully before attempting to release the tools.)

14. Avoid damaging the casing by pulling too much tension into the TAC.
15. Install the wellhead flange and set the landing slips and pack-off elements.

16. If using a flanged tubing bonnet for the tubing hanger, use low-profile rig slips to avoid pulling too much tension into the TAC (avoid damages to the old casing string). If using a threaded flange as a tubing head, you may lose some of the tension on the TAC when slacking off.

17. Install the flow line equipment and prepare to run the sucker rods.

Notes and Tips regarding Trips In or Out of the Hole with Production Tubing

- The tubing string must be checked for internal/external pits and damages.
- Keep the blocks at the center of the wellbore to avoid bending and thread galling.
- Avoid running the tubing string during high winds or weather storms.
- Protect the tubing from tong and slip damages.
- The tubing string must be scanned and the hydrostatic volume tested for leaks.
- Discard all tubing joints with bad threads.
- The pin and collars must be clean and free of rust, dust, mud, and foreign objects.
- Apply a light thread compound on the pin ends only.
- Make the tubing by hand for three rounds before applying hydraulic tongs.
- Avoid galling pins and collars to prevent tubing failures.
- Make up the tubing as recommended by the tubing manufacturing company.
- Always run the production tubing in the well slowly to avoid sudden stops.
- Always drift the tubing using full-size rabbits.
- Discard all bad couplings and damaged tool joints.
- Avoid damaging collars while tripping in the hole (hitting the wellhead).
- Use stabbing guides if and when required.
- Never hit the couplings with a hammer (hammer blows will cause splits and leaks under hydrostatic load pressure).
- Discard any and all tubing joints with dents, bends, pits, and sharp threads.
- Avoid slip and tong damages to the production string.
- Use wipers while tripping in or coming out of the well to avoid dropping objects in the well. Use wipers to wipe off oil and water from the tubing string.
- Lay down all the tubing joints with scale buildups while drifting and testing.
- If the tubing threads sink deeper in the tubing collar, do not run it in the well.

Preparing to Run the Insert Pump on the Sucker Rods

Based on the tubing landing details and seating nipple depth, you can actually calculate and space out the number of sucker rods you may need to reach the seating nipple in the well (you can even space these out for the pony rods needed).

Running the Insert Pump on the Sucker Rods

- ✓ Carefully set the insert pump on a higher elevation and inspect (check the suction and vacuum using the stroke length on the pump). You may check the pump at the well location using a vacuum gauge.
- ✓ Check the pull rod (valve rod) length and compare it with the pumping unit stroke length.
- ✓ Always lift the insert pumps using a special clamp (lifting a long insert pump to run in a well without a special clamp will bend and cause deformation on the pull rod/valve rod and may damage the barrel).
- ✓ Install the special lifting pipe clamp at the wrench, flat on the neck of the pump barrel, and lift the pump with a chain using rig blocks straight up.

Then slack off in the well and latch onto the pump using rod elevators.

✓ Fill up a clean bucket or container with diesel fuel. Lower the pump into the bucket to check and fill up the pump chamber with clean diesel fuel before going into the well (priming the pump with diesel until the diesel pours out from the top of the pump).

✓ You may screw a 20' gas anchor below the pump as a dip tube

or gas anchor (use a pipe clamp to pick up the tube to avoid dropping the gas anchor tube into the hole).

✓ Make up a gas anchor at the end of the fluid pump. Pick up the entire pump with a gas anchor to the rig floor (check the tube to make sure it is clean, with the holes fully open).

- ✓ Test and prime up the insert fluid pump with clean diesel fluid (this will keep the pump clean, preventing trash from getting into the pump while going in the hole). The diesel will soften and remove paraffin and dry grease from the balls and seats.
- ✓ Lower the pump in the hole.
- ✓ Install a 4' pony rod with molded rod guides on top of the insert fluid pump (to center the plunger valve rod and keep it from bending/wobbling from side to side while reciprocating).
- ✓ Always lift the insert pumps using special pipe clamp to lift

and/or lay down the pumps (a small pipe clamp will be installed at the wrench, flat on top of the pump barrel, to lift the pump and avoid damages to the pull rod).

- ✓ Install weight bars above the pump as required. Weight bars are used as stiff connections and extra weight to make the valve rod and plunger fall faster without bending and buckling against the tubing wall (the number of k-bars is based on the rod string design). K-bars and heavyweight bars may damage the tubing string in crooked holes. K-bars must not be too big to choke the fluid passing through the pump. If you are using fiberglass rods, you may need as many as twenty or more weight bars in some wells (the fiber rods

must operate in tension during the reciprocating cycle).

✓ Trip in the hole with sucker rods (fiber rods or steel sucker rods as designed). The sucker rods are the most important component of a beam pumping system. The rod string design could be either straight (all the rods are the same size) or tapered (different sizes of rods are used to make up the string). The tapered rod string may be arranged using any of following rod combinations:

- $\frac{5}{8}"-\frac{1}{2}"$
- $\frac{5}{8}"-\frac{3}{4}"$
- $\frac{7}{8}"-\frac{3}{4}"$
- $1"-\frac{7}{8}"$
- $1\frac{1}{8}"-1"$
- $\frac{7}{8}"-\frac{3}{4}"-\frac{5}{8}"$
- $1"-\frac{7}{8}"-\frac{3}{4}"$
- $1\frac{1}{8}"-1"-\frac{7}{8}"$

✓ Make up the rods by hand or using a rod wrench first. Then use hydraulic tongs to torque up the rods as recommended by the rod manufacturing company.

- ✓ Use proper torque to make up each size of rod string, using displacement cards as recommended by the rod manufacturing company.
- ✓ Lubricate the pin ends with special rod lubricant provided by the pump supplier. If no lubricant is available, use an equal mixture of 40-weight oil and a corrosion inhibitor.
- ✓ Do not hammer on the rods for any and all practical reasons.
- ✓ Do not hammer on the rod boxes (use a friction wrench to back off the couplings).

✓ Discard any and all the rod boxes with hammer blows.

✓ If you have to beat on a rod coupling or a tubing collar, you must replace it with a new one.

✓ All the couplings dealt with hammer blows will fail in the well first.

✓ Check all the used rods while hanging in the derrick. Lay down all the stretched rods.

✓ Never force making up the pin into a rod box (will create galling).

✓ Wash, tap, and clean all rusted rod threads and/or box threads using kerosene.

✓ Never use the rods and rod boxes for the wrong application (using slim-hole rod boxes in deep wells will cause the rod boxes to break in half).

✓ Do not hit or slap the rods with the rod elevator (under the rod shoulder).

✓ The rod boxes must have sufficient clearance in the tubing string for fluid to pass. They must also have sufficient clearance to fish out the rods in case of emergency.

✓ Do not walk on the rods. The rods must be clean and free from mud, dirt, and paraffin.

✓ Clean up the sucker rods and discard any rod with corrosion pits, wear, or bend.

✓ Finish tripping in the hole with pump and rods. Avoid too much slack while wrenching the rods (the rods must be standing straight when making up to avoid galling).

✓ Continue going in the well with the pump and rods to the top of the seating nipple.

✓ Tag up on the seating nipple with the rods slowly to find the bottom and mark the rod. Do not repeat seating and unseating the pump (you will cut and damage the cups).

✓ Space out the rods with pony rods and the polished rod assembly.

✓ Respace and finally seat the pump with a polished rod (do not jar down hard on the seating nipple to seat a pump).

✓ Pick up and space off the bottom as required (to avoid bumping up or down).

✓ A rule of thumb on spacing steel sucker rods

off the bottom is 12″ to 18″ (depending on the pump stroke and the valve rod length).
✓ The pickup space for fiber rods is different than that for steel sucker rods. Fiber rods

must be hung with no compression. Space out the fiberglass rods according to the following fiberglass spacing formula:

Total distance off the bottom = $\dfrac{9 \times \text{footage of fiber rods}}{1{,}000} + \dfrac{2 \times \text{seating nipple depth}}{1{,}000}$

✓ Rig up the horse head. Always use blocks and a sucker rod with a heavy-duty chain to lift and install the horse head on the beam (do not use skinny cat-line rope).

✓ Check the space off the bottom, pick up and install the polished rod clamp/clamps, and hang the rods on the wireline bridle/carrier (the carrier bar must be flat and level).

✓ Make up the stuffing box (change the rubber packing in the stuffing box if necessary)
✓ Install safety guards around the pumping unit to avoid serious accidents.
✓ Clear the guy lines and equipment and stay clear before starting the pumping unit (see the accident picture).
✓ Put the well on beam pumping mode. Check the unit for counterbalancing (rod heavy or weight heavy). Adjust the counterbalance if necessary to obtain smooth balance.
✓ Check the power surge on the up stroke and down stroke to save energy and to avoid burning the fuses, electric motor, and rubber belts.
✓ Lubricate the stuffing box and wait on the well to pump up. If the fluid is too low in the well, fill up the tubing with water to avoid burning the packing elements and to avoid standby time and rig costs.
✓ Filling up and testing the tubing string will indicate whether the pump is seated correctly. Filling up the tubing string may also indicate the fluid level in the well.
✓ Check and test the pump action and flow line leaks.
✓ Rig down the workover rig. Pick up all the trash and clean up the location.
✓ Turn the well over to operator to test and report workover results.

Running and Retrieving Tubing Pumps

(Pay attention to the tubing pump installation procedure.)

The name "tubing pump" is derived from the fact that the heavy-walled pump barrel is directly made up/screwed on the tubing string (the pump barrel and seating nipple become part of production tubing string). Tubing pumps are designed to produce a greater volume of fluid than standard insert pumps because of the larger bore sizes. The tubing pump consists of three distinct and separate components: the working barrel, the seating nipple, and the standard valve. These are directly made up on the production tubing string like a joint of tubing (become components of the tubing string).

The working barrel is a long heavy-walled tube of various sizes and lengths. It is made of either steel alloy or bronze (as strong as the tubing). The working barrel may be called the tubing pump because it is run on the production tubing string or may be referred to as the working barrel. A short standard API seating nipple or seating shoe is run at the end of the working barrel.

The seating nipple on the tubing pump is normally made up or screwed at the bottom of the working barrel in the pump shop and is delivered to a well location as one piece before it is run into the well. The seating nipple used on the tubing pump is a standard API nipple that is threaded externally on both ends and threaded internally at the bottom end for the gas anchor or dip tube. The seating nipple is polished and beveled on both ends.

Pump Plunger Assembly

The pump plunger with a traveling valve

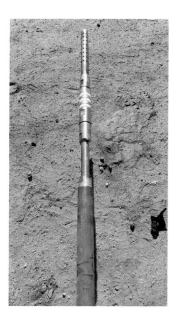

and puller tap is run on the sucker rod string. The traveling valve on the pump plunger is the discharge valve. At the bottom of the pump plunger, there is a small retrieving tool called the puller tap assembly, which consists of a small steel cage with a spring-loaded threaded tap (peanut) made up at the bottom of the pump plunger assembly. (The purpose of the pin tap is to screw in, catch, and retrieve the standing valve if and when necessary.) The peanut has only four rounds of small threads that can be pushed in and out by hand (poor design; need to upgrade the tool).

Standing Valve

The standing valve is a separate short piece by itself and consists of plastic nylon seating cups and/or composite ring cups with loaded cages of balls and seats. There are two seating cups without a no-go on the standing valve. The standing valve measures the fluid intake and acts as a check valve. It is an important part of the working barrel.

The standing valve may be run and seated in the seating nipple using three methods:

<u>First Method</u>

- The standing valve may be seated into the seating nipple at the bottom of the working barrel at the pump shop before running the production tubing string and working barrel into the well. If the standing valve is seated into the seating nipple/ seating shoe in the pump shop, you may be able to install a 20'- or 24'-long gas anchor below the standing valve before making up the working barrel onto the production tubing string.

<u>Second Method</u>

- You may drop the standing valve into the tubing string from the ground before running the pump plunger on the sucker rods. (You may use a 6" strainer nipple only below the standing valve if you did not run a gas anchor on the seating nipple. Strainer nipples will avoid trash getting into the pump.)

- Make sure there is no other gas anchor screwed below the seating nipple on the working barrel. (You cannot run and seat the standing valve using two gas anchors.)

- The tubing string should be full of fluid before dropping the standing valve down the tubing string. A high fluid level will slow down the free fall standing valve and prevent damages to the standing valve cups and the strainer nipple

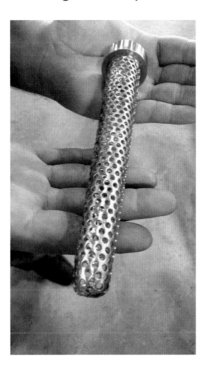

below it.

Third Method

- You may screw the standing valve at the bottom of the pump plunger loosely and run the pump plunger on the sucker rods (the standing valve is designed to be screwed onto the retrieving tap loosely).
- Do not screw on the standing valve too tightly. It will be difficult to back off a tight standing valve from the plunger.
- Once the tool reaches the bottom with the sucker rods, slack off and barely tag the bottom while checking the landing depth.
- Slack off and seat the standing valve first and then pick up while turning the rods to the left to back off from the standing valve.
- Turn the rods to the left using a hand wrench for just a few rounds and pick up to avoid pulling the standing valve out of the seat (do not apply hydraulic tongs).
- If you did not run a long dip tube below the working barrel, you may also run a 20' (correct size) dip tube below the standing valve and run the standing valve on the plunger only.
- Do not drop the standing valve with a long dip tube in the hole (keep the hole full before going in the hole).

How to Run a Working Barrel/Tubing Pump Assembly

Tubing Landing Procedure (From the Bottom Up)

i — Production tubing string (2 ⅜" or 3 ½" based on the casing size)
h — Full open TAC (must have the same ID as the tubing string)
g — 6' IPC pup joint (screwed above the working barrel)
f — Working barrel/tubing pump barrel (various sizes and lengths)
e — Seating nipple (already screwed below the working barrel)

d — Long gas anchor or dip tubing (20' or 24') (screwed below the seating nipple)
c — Perforated sub (4' or 8')
b — Two mud anchor tubing joints (2 ⅜", 2 ⅞" or 3 ½" based on the casing size)
a — Blind bull plug

Note that if you run the TAC above the tubing pump, the TAC must be fully open like the tubing string for the plunger assembly to go through it.

Tips

Just a reminder—if the standing valve is seated in the tubing pump prior to running the plunger with rods, you may expect the following:

- ✓ It may be difficult to swab the well if the fluid level is high in the tubing string. The swab cups may or may not fall through (the standing valve ball and seat are closed).
- ✓ You may not be able to pump fluid down the tubing string to circulate.

KHOSROW M. HADIPOUR

✓ You may be able to circulate the well down the annulus if necessary to remove trash in the standing valve.
✓ You will be able to long-stroke the pump to take trash out of the standing valve if necessary.
✓ Do not beat on the standing valve. You will damage the pin tap threads.

If you plan to run a TAC below a working barrel, you must use a steel tubing pump to avoid twisting off the brass barrel in case of an emergency or to release a stuck TAC.

Practical Procedure to Run the Tubing Pumps

1. Rig up the tubing tester and prepare to test the production tubing string and working barrel into the well if necessary (low test pressure will be applied to the brass pump barrel only).
2. The hydrostatic testing of a production tubing string is always a good practice. Testing the bars will drift the tubing, find tight spots, detect tool joint leaks and holes, and wipe off the tubing from the inside.
3. If the standing valve is already installed in the seating nipple below the pump barrel, you may have a chance to install a long dip tube or gas anchor below the standing valve.
4. Trip in the hole with two joints of tubing as mud anchor joints with a blind bull plug at the bottom of the joints (as listed before). (Mud anchor joints are useful and will accumulate mud, scale, sand, trash, and sludge that may otherwise get into the pump assembly.)
5. Avoid using oversized mud anchor joints to cause choking in the well (tubing collars).
6. Install a (4' or 6') perforated nipple on top of the mud anchor joints. The perforated nipple causes fluid intake, allowing fluid to enter the tubing string and the pump. The perforated sub may help direct gas interference and keep large trash from getting into the pump.
7. Check the seating nipple/seating shoe below the working barrel for internal damage. The seating nipple or seating shoe used on the working barrel is a heavy-walled nipple with external threads on both ends (we could use internal threads at the bottom as well).
8. Apply a light thread compound evenly on the pin ends of the tubing only.
9. If the standing valve is already seated in the seating nipple below the working barrel, you may be able to use a 20' dip tube or gas anchor at the end of the standing valve. The application of a gas anchor is useful to slow down gas locking in the pump. If using a long dip tube below the seating nipple, you may not be able to use a strainer below the standing valve any longer.

10. Pick up and lower the dip tube through the perforated sub and mud anchor joints. Then screw the small dip tube at the bottom of the seated standing valve in the working barrel assembly.

11. Make up the 6' coated pup joint on top of the tubing pump, Pick up and make up the barrel assembly onto the perforated sub (make up the barrel by hand first; then use tongs to torque up). Avoid using tongs or a pipe wrench on the brass tubing pump barrel.

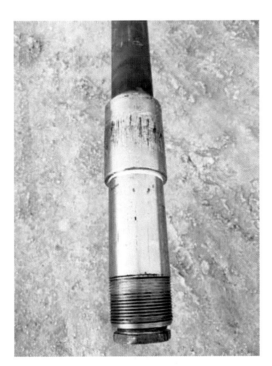

12. Lower the working barrel into the well (it is not necessary to do a hydrostatic test on the working barrel in the field; the tubing pump must be tested in the pump shop).

13. Install the TAC onto the pup joint above the working barrel (use double bow spring blades on the TAC as required).

14. The TAC must be fully open to allow the pump plunger to go through the pump barrel. If the tubing anchor has a smaller ID, then the TAC must be installed below the working barrel. The TAC must also be installed above any brass/bronze working barrel to avoid twisting off the soft elements on the working barrel in case of emergency (you may choose to use a steel working barrel if you decide to set the TAC below the working barrel).

15. Run and test the rest of the tubing string above the TAC. Make up all the tubing strings in three rounds by hand first; then use hydraulic power tongs to make up and torque the pipe as recommended by the manufacturer. Continue hydrostatic testing on the tubing as required.

16. Make sure the tubing anchor mandrel ID is fully open for the pump plunger to go through (the TAC must be fully open and have the same ID as the tubing string).

17. The TAC may be used below any steel working barrel only.

18. Finish going into the hole with the tubing string and working barrel assembly to the required landing depth (the pump depth).

19. Check the wellbore for flow. If there is no fluid flow, nipple down and remove the BOP.
20. Install and flange up the wellhead equipment.
21. Measure and set the TAC in tension at the required depth.
22. Pull and leave 7,000–8000 lb. tension on the tubing anchor only. (The required tubing tension is based on the tubing landing depth as well as the size and age of the casing string.) Excess tension on the tubing string may cause slip damage in an old casing string.
23. Avoid damaging the casing by over-pulling onto the TAC (if using a flanged wellhead, do not use long profile tubing slips).
24. Install the flanged wellhead and set the landing slips and pack-off elements, if any.
25. If using a threaded tubing flange for the tubing hanger, use low-profile rig slips to avoid pulling too much tension onto the TAC (will cause holes in old casing or parting tubing strings).
26. If running and landing the tubing is successful, prepare to run the plunger and rod string.
27. Rig up the floor and handling tools

to run the pump plunger and sucker rods.

Preparation to Run the Pump Plunger on Sucker Rods

- If the standing valve is already run with the working barrel, continue in the hole with the plunger and rods. If the standing valve is not run with the working barrel, go to the next step.

- Fill up the tubing string with water, if possible, before dropping the standing valve.
- Check the threads on the retrieving tap and drop the standing valve down the tubing string (do not apply any grease or pipe dope on the standing valve seating cups).
- If the fluid level is too low in the well, you may consider running the standing valve on the pump plunger. Loosely screw the standing valve at the bottom of the pump plunger assembly (use the screws on the tap threads) by hand and trip in the hole with a plunger and rods.

Preparation to Run the Pump Plunger and Sucker Rods

- Check the retriever tap at the bottom of the pump plunger (make sure the starting threads are not damaged or appear to be flat). Check the pump plunger for abrasions, cuts, and dents.
- Make up the plunger on 2′ of the sucker rod pony rod (do not use any rod guides on the pony rod).
- Never use rod guides right above the pump plunger assembly that might get into the working barrel.
- Do not use large weight bars on the pump plunger. (Large bars will damage the chrome inside the working barrel and cause abrasions.)
- You may use ⅞″ rods, 1″ rods, or 1 ¼″ weight bars without any type of rod guides above the pump plunger depending on the tubing and working barrel size (avoid using any rod guides that may get into the working barrel).
- Trip in the hole with the plunger assembly on the sucker rods. Lubricate the sucker rod threads and make up the rods by hand or using a rod wrench first; then apply hydraulic tongs to make up the rods. Use displacement cards for the appropriate sucker rod size and makeup torque.
- If you plan to use sucker rod guides to prevent rod and tubing wear

and/or cut paraffin, use three molded rod guides on each steel sucker rod only to reduce friction wear in the tubing.

- Use one rod guide 3'

above the rod box, one rod guide in the middle of the sucker rod, and one rod guide 3' below the upper box (do not choke the well by using large guides). I prefer molded-on rod guides.

Application of Sucker Rod Guides in Oil Well Beam Pumping

➢ People come up with all types of incomplete yet expensive solutions.
➢ The application of tubing rollers is an option in deviated holes with paraffin.

Sucker rod rollers

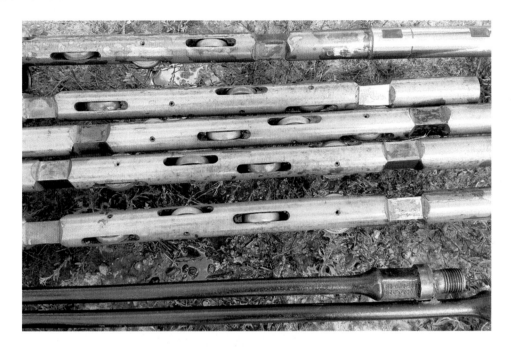

may fail because of thick paraffin and solids. They wear down on the low flat side and will stop rolling (only sliding up and down with the rods). You cannot use rod rotators on a well with rollers.

➢ Bladed and spiral steel paraffin cutter

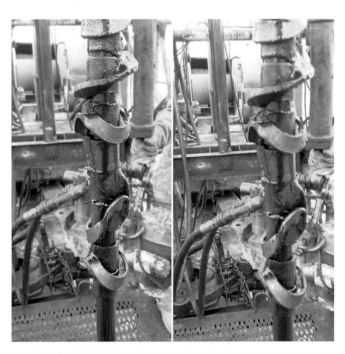

rod guides are actually hinged onto the sucker rods. They become stuck with heavy paraffin and sand. It is very costly to fish out stuck steel bladed rod guides out of tubing. They tend to slide and break down because of friction.

There are basically several types of rod guides sold in the market. The selection of good rod guides is based on the well and experience of the user.

➤ Manufactured molded-on rod guides

are semi-hard and stable during sucker rod strokes. These rod guides will not slide if they are applied correctly. The guides may wear down before the rod's life is spent (molded rod guides are fairly dependable products to use).

➤ Hard plastic rod guides are made of two pieces that are interconnected and made up around the rod body. These rod guides are normally easy to install on the sucker rods. These rod guides will become brittle with time and temperature and may break apart because of tight spots, crooked wells, and gas pounding (they slide on top of one another). Rod guides create expensive fishing work when they break around the sucker rods (very difficult to strip and fish sucker rods with loose junk rod guides).

➤ Plastic-type split rod guides will slide down on top of one another on up strokes and down strokes. The application of this type guide is useless and a waste of money. They fall on the rod box shoulder and expand, keeping the rods from going down (see for yourself).

➤ The snap-on rubber rod guide with a molded-on steel liner snaps on the sucker rods tightly. The method of snapping is by either a hammer blow force and/or pipe wrench that snaps the guide on the rod. Each method of rod guide installation is incorrect and may cause damage to the rod guides, causing pipe wrench marks or cracks on the rod guides.

➢ Snap-on rod guides will crack and become hard when exposed to temperature and gaseous wellbores. The penetration of gas will cause the rubber to balloon up and burst.

Improper installation and incorrect applications will cause the guides to slide down on top of one another. Fishing broken rod guides over rod boxes becomes difficult. The rubber-type snap-on rod guides will crack and swell up because of gas penetration. They wear off quickly. You may often see three or four rod guides on top of one another at the bottom of a sucker rod. Broken pieces of rubber will fall down, hang, and part the rods (stuck rods.)

- Continue in the hole with the pump plunger, weight bars, and sucker rods.
- Slow down three rods above the working barrel and seating nipple depth. Slack off and barely tag the bottom at the working barrel (you may see the plunger going into the tubing barrel because of small tolerance).
- Check and count the rods for the correct tagging depth. Slack off and reseat the standing valve.
- If the standing valve is run on the plunger, slack off and seat the pump. Pick up off the bottom while rotating the rods to the left (counterclockwise to release from the standing valve).
- Reseat the standing valve if necessary (tag and pick up with left-hand rotation).
- Mark the rods and space them out with pony rods and a polished rod.
- Space out the plunger off the bottom 18" or as required.
- Remove the rig floor and handling tools.
- Respace the rods off the bottom as required (fiber rods or steel rods must be spaced off the bottom to avoid pounding).
- Fiber rods must be hung in tension at all times to avoid compression and breaking the fiber rods (use fiberglass calculation for spacing).

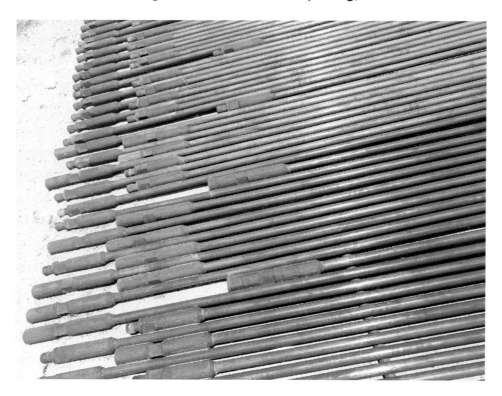

Total off-bottom distance = $\dfrac{9 \times \text{footage length of fiber rods}}{1,000} + \dfrac{2 \times \text{seating nipple depth}}{1,000}$

- Pick up the stuffing box and install it on the polished rod with the polished rod clamp.

- Make up the polished rod using a polished rod box only on all the steel sucker rods.

- Make up the stuffing box. Respace off the bottom again if necessary.
- Check and lubricate the wire rope "bridle" on the horse head with lube oil properly.
- Pick up and install the horse head (pick up the horse head with a chain and blocks).
- Make sure the carrier bar is hung flat and level to avoid breaking the polished rod while reciprocating.
- Clear the area and move off all the guy lines away from the horse head.
- Put the well on pumping mode (walk around the unit to check for abnormal loose parts).

- Check the up stroke and down stroke (make sure the rods are not tapping down or up; the pounding bottom will damage the standing valve, and tapping on the up stroke will damage the surface equipment).

Causes of Rods Tapping on the Up Stroke or Down Stroke

➢ Incorrect rod spacing

➢ Loose objects around the rods or plunger
➢ Gas locking

- Respace if necessary (bumping the bottom will damage the retriever tap and peanut threads, and you may not be able to retrieve the standing valve later).

- Wait on the well to pump up (fill up the tubing with water if necessary).
- Rig down and release the rig. Clean up the location.
- Secure the pumping with safety guards. Move off location.

Tubing Pump Diagnostic Failures and Repairs

Because of the multiple failures of tubing pump components, it may be necessary to pull out the downhole equipment to find out the source of these problems.

- Parted tubing string
- Hole in the tubing string

- Bad standing valve or traveling valve
- Parted rods

- Sanded-up standing valve
- No fluid entry

• If the Sucker Rods Are Parted . . .

Test the tubing string first. Then fish out parted rods, replace bad rods, and check and repair plunger components. Run the rods and plunger back into the well. When fishing the sucker rods, it is not necessary to pull the standing valve and/or tubing string out of the well.

• If There Is a Hole in the Tubing String . . .

You must pull all the components of the tubing pump out (rods, plunger, standing valve, and tubing string). You must also pull the sucker rod string with the pump plunger and standing valve. You may have to screw into the standing valve assembly (retrieve the standing valve with the rods). At last, release the tubing anchor

and pull the entire production tubing out of the well to repair/replace bad tubing joints.

- How to Pull Rod String with Standing Valve

  - Slack off with the sucker rods and pump plunger to the top of the standing valve.
  - Rotate the sucker rods to the right screw into the standing valve.
  - Latch into and pick up to unseat and recover the standing valve.
  - Pooh with the rods, plunger, and standing valve (send the plunger and the standing valve for inspection and repair).
  - If you are unable to unseat the standing valve, prepare to pull the "wet" tubing string.
  - You may have to perforate the tubing to pull the tubing string.

- How to Pull Tubing String and Working Barrel (Tubing Pump)

- Kill the well and circulate the well if necessary. Nipple up and release the TAC. Pull the tubing string and working barrel out of the well.
- Check the entire tubing string with the tubing pump assembly (send the tubing pump to the pump shop for inspection and repairs).
- Check, repair, and replace the tubing string as required. Test the tubing and bottom hole assembly and rerun the tubing string and tubing pump in the well.
- Run the rods with a pump plunger and put the well on production.

- If You Have a Bad Standing Valve . . .

You may lower the rod string and plunger assembly. Screw into the standing valve and retrieve the valve on the surface. Check and repair the standing valve and run the rods with a plunger and standing valve. Seat the standing valve and put the well on production.

- If You Are Unable to Retrieve the Standing Valve . . .

You may have to circulate the wellbore clean and kill the well first. Pull the sucker rods with the pump plunger and then the tubing string. Pulling the entire tubing string requires pulling the "wet" tubing string and/or perforating the tubing string to avoid pulling the "wet" tubing string to recover the standing valve. Pulling a "wet" tubing string is always time-consuming, unsafe, and risky (it is your decision).

- **If You Suspect a Bad Traveling Valve on the Plunger . . .**

You must pull the sucker rods and repair or replace the valves in the plunger assembly only. You do not need to pull the standing valve or tubing string out of the hole.

- **Sanded-Up or Stuck Sucker Rods and Pump Plunger**

It is advised to evaluate the well problem with this complicated condition. Estimate the costs associated with the workover cost before moving a rig to the well.

- Strip out and fish the sucker rod string and pump plunger.
- Release the TAC.
- Fish and pull the tubing string out of the wellbore.
- Wash and clean up the wellbore and perforations.
- Check the casing for leaks and squeeze if necessary.
- Test and run the production equipment.
- Put the well back on production.

- **Wellbore Remedial Intervention**

All the production equipment must be pulled out (sucker rods, plunger, standing valve, and tubing string) to conduct any well interventions.

Procedure to Retrieve and Repair Tubing Pumps

The standing valve or the tubing string—you do not know which is leaking.

1. Move in and rig up the workover rig (service pulling unit).
2. Secure the guy lines to a base beam or permanent safety anchors on location.
3. Test the production tubing for leaks before unseating the standing valve.
4. If the tubing string fails to hold pressure, it could be a hole in the tubing string or standing valve leaks. In either scenario, the tubing must be pulled to the surface for repairs.
5. Bleed off the tubing and casing to zero pressure.
6. Circulate to kill the well if necessary (circulating fluid may clean up and remove solids above the standing valve).
7. Un-beam the pumping unit, pick up on the polished rod, and set the rod string on the stuffing box to remove the horse head.

8. Rig down the horse head and spot the head on the ground away from the rig crew. Use rig blocks to pick up or lay down the horse head (do not use cat-lines).

9. Use a heavy-duty chain to pick up and lay down the horse head off the beam (never use cat-lines to pick up or lay down heavyweight horse heads).

10. Check the well for flow.(bleed off the casing and tubing pressure).

11. Check the wellbore for H2S and inform the workover crew (safety first).

12. Kill the well with heavy water if necessary (the well fluid must be static during workover).

13. Pour five gallons of diesel down the tubing string to remove or soften paraffin.

14. Latch on the polished rod and back off the stuffing box (make sure there is no fluid flow).

15. Pull and lay down the polished rod and stuffing box and place them off the ground.

16. Avoid hammering the polished rod box when disconnecting the polished rod from the string.

17. Pick up an extra rod and slack off with the rod string to retrieve the seated standing valve.

18. Rotate the rods going down to the right to screw into the standing valve. Repeat several times if necessary to make sure the retrieving tap is engaged into the standing valve. Do not bump or set too much weight on the standing valve. Too much weight will push the retrieving threads back into the cage housing and will not catch the standing valve.

19. Pick up on the rods with right-hand torque to unseat the standing valve (you will see the standing valve pulling free, coming out of the seating nipple assembly with a slight vacuum on the tubing).

20. Pull out of the hole with rods and the standing valve. Lay down the plunger and standing valve.

21. If you are unable to screw into the standing valve, you may have any of the following downhole problems:

- Bad threads on the retrieving tool at the bottom of the plunger
- Sand, scale, or solids above the standing valve
- Deeper parted rods
- Cannot get deep enough to latch into the standing valve because of bad tubing
- Incorrect retrieving procedure

If you are unable to retrieve the standing valve, the tubing string must be pulled to resolve the well problems.

22. You are facing the risk of pulling the tubing while it is "wet" (safety issue). The tubing string must be perforated to avoid pulling the "wet" tubing string. If you decide to pull the tubing while it is "wet," you may need to plan for pollution control.
23. Nipple down the wellhead equipment and nipple up the BOP. Test the BOP at 200 psi (low) and 3,000 psi (high).
24. Rig up and prepare to scan the tubing string if necessary.
25. Latch on the tubing string, release the TAC, and pooh and scan the tubing string coming out of the hole if necessary. If there is no hole in the tubing string, you may have to pull the tubing string "wet" and/or perforated or swab the tubing to avoid pulling the "wet" tubing and avoid oil and salt water pollution. Note that pulling the tubing string "wet" with the standing valve in place is always unsafe.
26. Finish out the hole while laying down the red-band and green-band tubing. Stand back yellow-band and/or blue-band tubing. Lay down the working barrel and bottom hole assembly.
27. Watch for the well kick when laying down the working barrel and perforated nipple. You cannot shut the well in with the perforated nipple inside of the BOP stack. Send the tubing pump and plunger (insert to the pump shop for inspection and repairs.
28. Rig up the tubing testers, hydro test tubing string, and working barrel in the well.
29. Run the rod and plunger and put the well on pumping mode.

Note that when attempting to fish out and retrieve the standing valve with the sucker rods, you may or may not get it. After pulling the sucker rods, you may experience the following:

➢ The rods and pump plunger came out but did not recover the standing valve.
➢ The rods are parted deep in the hole (did not come out with the plunger or standing valve).

- If you just missed pulling the standing valve, you may trip back with the rods and a new retrieving tap tool to fish the sanding valve out (it is a cost-saving trip).
- If you find out that the sucker rods are parted deep in the hole, you must trip in the hole with the correct-sized overshot to fish out the parted rods and the plunger only.

- To repair a leak in the tubing string, the sucker rods, pump plunger, standing valve, and tubing string must be pulled out of the well.
- To repair the standing valve only, the standing valve must be pulled out of the well. You must lower the rods with a plunger, screw into the standing valve to fish it out, and retrieve the standing valve for repair.
- If you are unable to screw into the standing valve because of sand, solids, or galled threads, the entire tubing string must be pulled "wet" to recover the standing valve. The rig must lift the entire tubing string and its fluid contents (tubing, oil, water, gas, and solids in a deep well). Let the rig supervisor know the condition you are facing.

To Avoid Pulling the Tubing String "Wet" . . .

- The standing valve must be pulled out of the seating nipple.
- You must perforate a hole in the tubing string above the working barrel.
- Swab the fluid content from the tubing string into the casing or into a rig tank.
- The worst-case scenario is when the rods are parted deep. After you latch onto the rods, you may not be able to pull out the stuck pump and cannot release the overshot facing the stuck rods or stuck plunger. The workover operation conditions may become nasty, dirty, and unsafe.
- Another major problem that you may have to deal with is a stuck TAC, a hole in the tubing, and a stuck pump and/or sucker rods.

Running Orders of Tubing Pump Components

- <u>Blind Bull Plug</u>

This keeps the produced solids from falling out of the mud anchor joints.

- <u>Mud Anchor Joints</u>

This is used to store and hold mud, sand, and trash and keep them from getting into the pump. Avoid using oversized mud anchors that may choke the fluid into the pump.

- <u>Perforated Sub</u>

This is a short 4' pup joint with slots or round perforated holes. The purpose of the perforated sub is to allow fluid into the pump intake. On some pumping designs, you may have an option to leave the tubing open-ended. Using an 8' perforated sub in low-gas-producing wells is a good option.

- <u>Working Barrel with the Seating Nipple</u>

The working barrel is made of soft brass/bronze or steel and comes in various sizes. At the bottom of the working barrel is a steel seating nipple with eight-round male threads on both ends to screw onto the perforated sub and the bottom hole assembly.

- <u>Seating Nipple</u>

 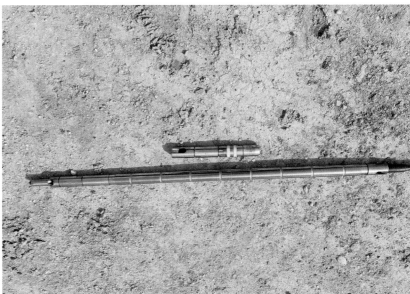

The seating nipple on the tubing pump is a short ex-heavy nipple tube that is threaded externally on both ends and also threaded internally at the bottom. The seating nipple must be free of corrosion pits, cuts, and rough spots internally. Make sure the seating nipple is an API-designed nipple. The seating nipple is designed to hold the seating cups and support the pump in place (non-API aftermarket seating nipples are available but may not hold the pump cups). There are various types of nipples available. Make sure you are using a correct-sized seating nipple on the tubing pump. The seating nipple on the tubing pump barrel is short in length and threaded on both sides on the outside as well as threaded internally at one end for the gas anchor.

- <u>Full Open 6′ Tubing Sub (Internal Coated Pup Joint)</u>

This sub is necessary to space out between the working barrel and the tubing string.

- TAC

The TAC is made to hold the tubing string in tension, keep the tubing from moving side-to-side, and catch the tubing string from falling in case of parted tubing.

- Production Tubing String

The tubing string is the most important part of any artificial lift system. Well production is lifted through the tubing string and up to the surface. The tubing string must be checked frequently for internal and external corrosion pits and damages.

> The pin and collars must be clean and free of sand, mud, and foreign objects.
> Apply a thread compound on the pin ends only before making up the pipe.
> Make the tubing by hand for three rounds before applying hydraulic tongs.
> The tubing pin must not sink deep into the tubing collar before makeup.
> Make up the tubing as recommended by the tubing manufacturing company.
> Always run the production tubing slowly in a well to avoid a sudden stop.
> Always drift the tubing with full-size rabbits.
> Discard all bad couplings and damaged tool joints (must use seamless coupling).
> Avoid pinching, cuts, and damaging to the tubing when replacing bad coupling.
> Use stabbing guides if required.
> Never beat on the couplings with a hammer.
> Use wipers while tripping in or coming out of the well.
> Avoid damaging the casing by over-pulling into the TAC (do not use tall slips).
> Install a wellhead flange and set the landing slips and pack-off elements.
> If you are using a tubing bonnet for the tubing hanger, use rated low-profile rig slips to avoid pulling too much tension into the TAC (this will avoid damages to the tubing and old casing).
> The standing valve may be made up and seated in the pump barrel at the pump shop, dropped through the tubing string, and/or screwed loosely to the retrieving tool below the pump plunger before running into the well (driving the standing valve is a choice).

- Tubing Pump Plungers

The tubing pump plungers are manufactured and designed similar to insert pump plungers. The plungers are either solid chrome-plated or ring/grooved plungers

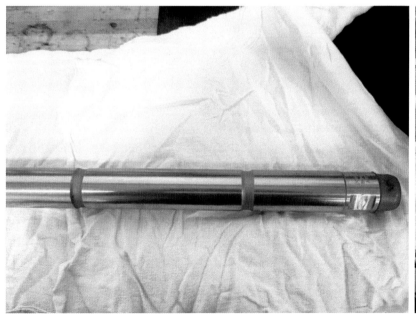
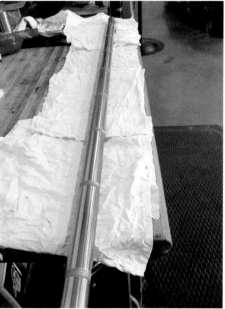

to handle moderate sand, solids, and more fluid. The distinct advantage of the tubing pump is its ability to pump large volumes of fluids out of the wellbore, and it may bypass sand or solids and gas better than insert pumps.

Principal Applications of Sucker Rods

Sucker rods do not suck at all. They are the major part of transferring energy from the ground to the subsurface pump and from the fluid pump up to the surface. Sucker rods are screwed together with sucker rod couplings. The rods are connected to the subsurface downhole pump and extended up the hole to the polished rod and the carrier bar at the horse head of a pumping unit above the ground. Sucker rods are important components of beam pump artificial lifting. They are the main component that connects the subsurface fluid pump to the surface polished rod.

Using Sucker Rods in Beam Pumping

To successfully reciprocate the pump plunger on a beam artificial lift and to displace the well fluid to the subsurface, one must have a durable and strong link between the pump plunger and the surface pumping unit. The connecting object must be stiff and strong enough to carry the heavy polished rod loads.

The Best Artificial Fluid Lifting Concepts for Beam Pumping

- Continuous steel pipe (similar to coiled tubing)

This lifting concept will work but requires too many well repairs because of corrosion and erosion.

- Steel wire rope

The concept of using steel wire ropes is similar to swabbing fluid out of a well. The wire rope is strong enough to pull heavy loads; however, steel rope will not be suitable for complete reciprocating cycles as well as downhole erosion and corrosive environments.

- Continuous sucker rod

The application of a continuous rod could be an ideal method of lifting fluid and saving costs; however, the costs of rod repair and fishing will outweigh the advantages of the idea (a continuous rod string was applied in the past but did not work quite well in the oil field).

- Fiber rods

Fiber rods are tough, stiff, straight, flexible, light, noncorrosive, and efficient enough for beam pumping operations. They will have practical applications in oil wells and artificial lift. A continuous fiber rod string may work, but this has not been tried yet.

- Steel sucker rods

Steel sucker rods are tough, stiff, and straight but heavy. They are susceptible to erosion and corrosive environments. Steel sucker rods will work well in the oil well beam pumping concept.

The types of sucker rods available in oil and gas operations are the following:

- Steel sucker rods of various sizes and grades
- Fiberglass sucker rods of various sizes and grades

Principal Applications of Steel Sucker Rods in Artificial Lifting

Steel sucker rods are long straight round bars that come in various sizes and grades. They are manufactured from high-quality hot-rolled carbon steel and carbon steel alloy bars, straightened, inspected, heat-treated, shot-blasted, and re-inspected for body surface defects and cracks.

The steel sucker rod consists of the following components:

- Threaded pin
- Pin shoulder
- Wrench flat (wrench squire)
- Bead
- Rod body (25'–30' long)

The sucker rod ends are hot-forged. As the result of continuous engineering research and efforts, sucker rods are built to last longer, with fatigue resistance and corrosion tolerance.

Rods are normalized from end to end, heat-treated, quenched, and tempered. The entire rod body should be shot-blasted and all the scale removed before the rod threads are cold-cut.

Steel bars may go through a long process:

- Raw iron material going through an electric furnace to turn into liquid steel
- From liquid steel to steel bars of various round shapes
- From round steel bars to the hot-rolled process to obtain different rod sizes (⅝", ¾", ⅞", 1", 1 ⅛").
- Forging process (to enhance resistance to fatigue stress)
- Complete heat treatment, quenching, and full-body normalizing to achieve the required mechanical properties
- Reheating and shot-peening to achieve resistance to fatigue and hardness
- Re-inspecting and testing before cold-rolling the threads
- Final inspection and electronic testing
- Surface coating
- Installing end protectors
- Storing
- Marketing and shipping to the oil field

Chemical Properties of Sucker Rods

The added elements in steel sucker rods will change the physical and/or chemical properties of the sucker rods. These added elements must be carefully evaluated with a testing balance.

The following elements may be added in the construction of steel sucker rods:

- Carbon (C) — Increases hardness, tensile strength, and abrasion resistance (the relation between carbon and steel can be described as "salt and water")
- Manganese (Mn) — Increases hardness and abrasion resistance and decreases ductility
- Silicon (SI) — Increases hardness, tensile strength, and deoxidization
- Aluminum (Al) — Increases deoxidization and strain aging
- Copper (Cu)
- Molybdenum (Mo)
- Chromium Cr) — Increases hardness and corrosion resistance
- Nickel (Ni)
- Phosphorous (P) — Increases tensile strength and hardness, decreases ductility, and improves machinability
- Sulfur — Increases, machinability, hardness, and corrosion resistance and decreases weldability, ductility, and toughness

Other elements may be added to steel to improve its mechanical strength and resistance against corrosive environments (vanadium, molybdenum, nickel). Each of the above elements will have a positive effect on the physical and chemical conditions, tensile strength, hardness, ductility, and resistance to corrosion in steel rods.

The demand for good-quality sucker rods is enormously high in artificial lifting. The sucker rod's quality has not met operation expectations yet. Check to ensure that the rods are not made of junk cars and surplus material. No one knows where the sucker rods of today come from.

Normalizing, tempering, and quenching will result in high strength and higher carrying loads on sucker rods (higher tensile strength). The acceptable length of new steel sucker rods in the oil field may range from 25' to 30' according to the API. The most available and practical length of steel sucker rods in the oil field is 25'. These rods may be run or pulled in singles and/or triples (depending on the availability of the rig mast).

Steel sucker rods are divided into three grades:

- Grade "C" (shallow depth)
- Grade "K" (shallow to mid-range depth)
- Grade "D" (deeper-range wells)

Extra-strength rods are available in the market today. The misapplication of rod grades will results in rod string failure. The 25'-long steel sucker rods of today are available in these sizes:

Rod OD Size	Dry Rod (weight/foot)
½"	0.68
⅝"	1.14
¾"	1.63
⅞"	2.22
1"	2.90
1 ⅛"	3.67

Rod couplings are used to connect the rods together. There are two types of sucker rod couplings:

1. API-class "T" coupling — This coupling is cheaper and appears without hard chrome-plated surface areas. It also appears to be softer and has less resistance against high H2S and corrosive fluids. You may back off the T coupling using a pipe wrench (soft metal).
2. Hard chrome alloy surface coupling — This coupling is tough, with high resistance against corrosion and erosion. It is either a slim hole or a full hole. Hard surface spray couplings will outlast the sucker rods. It is difficult to back off the box from

the rod without damaging the coupling. The hard surface rod box is referred to as the spry box (spry coupling).

The spry box coupling is more costly but outlasts the useful lives of sucker rods and the tubing string. When you changing the rod string, you will see that the spry couplings are still in very good condition. (Save the spry boxes using friction wrenches. Do not waste expensive rod boxes to slow down tubing wear.) Spry couplings may drastically increase rod wear in the tubing string in deviated and crooked wellbores (molded rod guides are recommended above and below the boxes).

Rod Coupling Sizes

	Full-Size Box	Slim-Hole Coupling	Length
⅝" box = 0.625"	1 ½"	1 ¼"	4"
¾" box = 0.750"	1 ⅝"	1 ½"	4"
⅞" box = 0.875"	1 13/16"	1 ⅝"	4"
1" box = 1.00"	2 3/16"	2"	4"
1 ⅛" box = 1.125"	2 ⅜"	2 ¼"	4 ½"

Names and Dimensions of Steel Sucker Rods

Oil field steel sucker rods are manufactured to be 25' to 30' in length, with threaded pins on both ends without rod couplings (rod couplings are separate pieces that you have to buy with the rods). In the old days, the sucker rod was designed with a threaded pin on one end and the rod coupling that was part of the rod on the other end (this design idea did not last too long because of coupling wear). The rod coupling was actually a permanent part of the sucker rod. The steel sucker rod consists of several sections: the threaded pin, the pin shoulder, the wrench flat (wrench square), the bead, and the plain rod body.

Pony Rods (Short Subs)

Pony sucker rods are shorter in length. The purpose of the short pony sucker rod is to space the rod string below the polished rod to avoid the rods from bumping on the up stroke or down stroke. The pony rod may be added above a fluid pump and/or below the polished rod assembly for spacing. Pony rods are stronger than full-length sucker rods of the same size.

Pony rods are available in the following lengths and sizes:

Length	Size
2'	($\frac{5}{8}$", $\frac{3}{4}$", $\frac{7}{8}$", 1", 1 $\frac{1}{8}$")
4'	($\frac{5}{8}$", $\frac{3}{4}$", $\frac{7}{8}$", 1", 1 $\frac{1}{8}$")
6'	($\frac{5}{8}$", $\frac{3}{4}$", $\frac{7}{8}$", 1", 1 $\frac{1}{8}$")
8'	($\frac{5}{8}$", $\frac{3}{4}$", $\frac{7}{8}$", 1", 1 $\frac{1}{8}$")
10'	($\frac{5}{8}$", $\frac{3}{4}$", $\frac{7}{8}$", 1", 1 $\frac{1}{8}$")

In the oil field operation, the following rod strings may be designed and run:

   a) Straight rod design (all the rods run in the well will be the same size)
   b) Tapered rod string (consists of two or three sizes of sucker rods)

The tapered rod string is usually run in a deeper well to reduce the pick polished rod load and to reduce costs.

Sucker Rod Overshot and Fishing Tools in the Oil Field

Fishing and Safe Pulling on the Steel Sucker Rods

Check the rod size and rod grade before pulling into the rod string. (Avoid permanent rod string damage.) The safety factor for pulling onto a rod string should be considered to prevent breaking the rods or causing permanent rod stretch damage. For "new" steel sucker rods, the SF is equal to 90% of the minimum yield strength. For used steel sucker rods, the SF is equal to 70% of the minimum yield strength.

a) The straight rod string design means that all the rods are the same size.

Safe pulling = (rod cross-section area) × (minimum yield) × (safety factor)

b) The pulling load on a tapered rod string is based on the smallest diameter of the rods in the rod string.

Safe pulling = (cross-section area of the smallest rod string) × (minimum yield strength) + (the weight of all the rods above the smallest rod string)

Note that pulling the load using minimum yield strength without the calculated safety factor is not correct and will not be recommended as a safe pulling practice (the yield strength of old sucker rods is different from that of new sucker rods).

Rod Size	Area	Minimum Yield (New Rod) Minimum Yield Strength × .90		Minimum Yield (Used Rod) Minimum Yield Strength × .70	
		C & K Grades	D Grade	C & K Grades	D Grade
⅝"	0.308	26,000	31,000	19,000	24,000
¾"	0.442	37,000	45,000	28,000	35,000
⅞"	0.601	51,000	60,000	38,000	48,000
1"	0.785	66,000	80,000	50,000	61,000
1 ⅛"	0.994	82,000	100,000	66,000	80,000

Sucker rods may be run in a well as straight one-size rod strings in shallow wells (¾", ⅞", 1", or 1 ⅛"; all the rods will be the same size and the same grades only); 1" and 1 ⅛" rods may not be run inside of 2 ⅜" tubing because of drift diameter, while 2 ⅜" ID (1.995") and 1" rod boxes (2") cannot be fished out. Sucker rods may be designed as tapered rod strings (double- or triple-rod strings)—(¾", ⅞"), (¾", ⅞", 1"), or any other calculated combinations.

Fishing Tools (Baby Red Overshot)

Stop wild-pulling tension without calculations. Calculate the pulling load first. Then slow the upstream lifting movement to avoid permanent deformation (avoid sharp jarring or bumping to exceed minimum yield strength).

Rod Coupling OD Size

Rod Size	Slime Hole Box OD	Full Hole Size Box OD
⅝"	1 ¼"	1 ½"
¾"	1 ½"	1 ⅝"
⅞"	1 ⅝"	1 13/16"
1"	2"	2 3/16"
1 ⅛"	2 ¼"	2 ⅜"

Parted Rod Size	Fishing Tool Size
⅝" pin break	1 7/16" split socket or snap ring
⅝" rod box	1 ⅜" split socket or snap ring
⅝" worn box	1 7/16" sleeve
⅝" SH pin or box	Takes a special tool (baby red)
⅝" body break	Special overshot ⅝" × ¾" slips
¾" SH box breaks	1 ½" overshot tools (1 9/16")
¾" worn box	1 9/16" sleeve
¾" reg. box break	1 ⅝"
¾" worn pin break	1 7/16" sleeve
¾" pin break	1 ½" sleeve
⅞" worn box	1 ¾" sleeve
⅞" reg. FH	1 13/16"
⅞" SH box	1 ⅝" or 1 9/16" tool
⅞" worn pin	1 9/16"
⅞" pin	1 ⅝" tool
⅞" reg. box	1 13/16"
⅞" body break	¾" × ⅞" slips
⅞" full-hole box in 2 ⅜"	Cannot be fished
1" body break	1" slips
1" box	2" sleeve
1" worn box	1 15/16" or 1 ⅞" sleeve
1" SH box break	2" tool
1" pin break	2" tool
1" reg. box break	2 ⅛" tool
1" reg. box in 2 ⅜"	Cannot be fished
⅝" and ¾" body break	Will take the same slip socket to fish
⅞" body break	Will take the same slip socket as ¾" to fish
1" body break	Will take a 1" slip socket to fish

Steel Rod Dimensions

⅝" steel rod = 0.625" (1.14 lbs/ft)
⅝" thread diameter (15/16") = 0.9375"
⅝" SH box = 1 ¼" × 4" long
⅝" FH box = 1 ½" × 4" long

Rod Dimensions

¾" rod body = 0.75"
¾" rod shoulder = 1 ¼"
¾" pin shoulder = 1 1/16"
¾" pin size = 1 1/16" × 1 ½" long
¾" box FH = 1 ⅝"
¾" box SH = 1 ½"

Rod Dimensions

⅞" sucker rod = 0.875" (2.22 lbs/ft)
⅞" rod shoulder = 1 ⅜"
⅞" pin shoulder = 1 3/16" × 1 ⅝" long
⅞" SH box = 1 ⅝" OD
⅞" FH box = 1 13/16"

Rod Dimensions

1" sucker rod = 1"
1" rod box SH = 2" OD
1" rod box FH = 2 3/16"
1" pin shoulder = 2"
1" rod shoulder = 1 ¾"
1" pin size = 1.3"

Tapered Rod String Components (Length and Design)

Rod	¾"–⅝"	⅞"–¾"	1"–⅞"	1⅛"–1"	¾"–⅝"–½"		⅞"–¾"–⅝"		1"–⅞"–¾"		1⅛"–1"–⅞"	
Pump Size	% ¾"	% ⅞"	% 1"	% 1⅛"	% ¾"	% ⅝"	% ⅞"	% ¾"	% 1"	% ⅞"	% 1⅛"	% 1"
1 1/16"	34.5	28.6	24.4	21.2	33.4	33.1	27.2	27.4	22.6	23	19.7	20
1 ¼"	37.4	30.6	25.8	22.3	37.2	35.9	29.4	29.8	24.3	24.5	20.8	21.2
1 ½"	41.8	33.8	27.7	23.8	42.3	40.4	33.3	33.3	26.8	27	22.5	23
1 ¾"	46.9	37.5	30.3	25.7	47.4	45.2	37.8	37	29.4	30	24.5	25
2"	52	41.7	33.3	27.7	0	0	42.4	41.3	32.8	33.2	26.8	27.4
2 ¼"	0	46.5	36.4	30.2	0	0	46.9	45.3	36.9	36	29.4	30.2
2 ½"		50.9	39.9	32.7					40.6	39.7	32.5	33.1
2 ¾"		56.6	43.9	35.6					44.6	43.3	35.1	35.3
3 ¼"			51.6	42.2						42.9	41.9	
3 ¾"				49.8								

Rod Coupling OD Size

Rod Size	Slime Hole Box OD	Full Hole Size Box OD
⅝"	1 ¼"	1 ½"
¾"	1 ½"	1 ⅝"
⅞"	1 ⅝"	1 13/16"
1"	2"	2 3/16"
1 ⅛"	2 ¼"	2 ⅜"

Use spry couplings in wells with H2S corrosion problems.

Safe Pulling on Tapered Sucker Rod String

Always consider the small rod string (¾", ⅞", 1" tapered rod string).

Safe pull (SPL) = (cross-section area of small rod) × (minimum yield strength value of the smallest rod string) × (safety factor of the smallest rods) + (weight of the rod string above the smallest rod string section)

The steel sucker rods are classified in the following grades:

- Grade C rods — Grade C steel sucker rods and pony subs are good for shallow depths with light loads only.

- Grade K rods — Grade K rods and pony rods are good for light loads in shallow to mid-range well depths.

- Grade D rods — Grade D carbon steel rods are designed to perform in shallow as well as deep wells of moderate to heavy loads. Grade D rods will have higher yield strength and tensile strength to perform in deeper wells with noncorrosive fluid.

Other types of high-strength series of rods are available in the market. Always compare steel suckers manufactured with different companies and select the best rods for the wellbore conditions. The grades of steel sucker rods denote the mechanical and chemical characteristics of the manufactured steel sucker rods for specific downhole applications. Make sure you do not run mixed grades of rods in deep wells. Always read the manufacturing label on the wrench flat to identify the type of rod you are using.

Tensile strength and corrosion resistance are important trades in steel sucker rod selection. High-strength steel alloy rods are used in deeper well applications; however, high-strength steel rods with high tensile strength may be susceptible to downhole H_2S and CO_2 corrosion. Nearly all the wells in the United States will contain H_2S and CO_2. High-carbon-content rods are also susceptible to pits and cracks because of corrosion.

Wellbore Evaluation before Rod Selection

- Pumping depth
- Pumping speed (strokes per minute)
- Pump size
- Fluid volume

- Corrosive fluid such as CO_2 and H_2S
- Friction sources, erosion, deviation, sand erosion, and paraffin problems
- Tubing size
- Fluid density
- Oil/water ratio
- One-size rod string design vs. tapered rod string design

Steel sucker rods are shipped in standard pallets:

⅝"	120 rods in a pallet
¾"	100 rods in a pallet
⅞"	80 rods in a pallet
1"	60 rods in a pallet

Sucker rods may be designed as straight one-size rod strings (⅝", ¾", ⅞", 1", or 1 ⅛"). Sucker rods may be designed as tapered rod strings (double or triple rod strings)—(¾", ⅞"), (¾", ⅞", 1"), or any other calculated combinations. A good tapered rod string design will eliminate unnecessary heavy dynamic loads in deeper wells and avoid rod stretch, rod failures, polished rod fatigue, and overloading the pumping unit gear box.

Why Sucker Rods Fail

The enemies of oil field sucker rods are too many. Sucker rods are too weak for the work they perform. It is a major challenge for oil and gas companies to keep up with rod failures, one after another. The causes of rod failure are too many. It is very difficult to keep up with downhole changes while operating a string of sucker rods for months without the high cost of maintenance. The downhole and well equipment must be kept in good working condition to protect the sucker rods from failing. Often rods may break because of multiple problems that cause the failures.

The top reasons behind major sucker rod failures are as follows:

- Corrosion attacks (e.g., H_2S, CO_2, O_2)
- Human carelessness in rod handling
- Downhole conditions(gas pound, and water hammering)
- Poor rod selection and rod design
- Pumping speed (over travel and under travel)
- Unclean wellbore

- Fatigue failures are due to fast pumping speeds (reciprocations per minute). Small sucker rods, fast rod strokes, and heavy polished rod loads are direct causes of fatigue failure (clean, smooth body breaks). One of the major reasons that sucker rods fail is the change of load. Slow pump strokes with a long stroke length are recommended to prolong the useful life of sucker rods.

• Corrosion is the deterioration of an element in contact with the environment. It is known to be the worst enemy of steel sucker rods,

causing pits, cracks, worm grooves, and stress fatigue. Through proper care and the consistent application of compatible chemical inhibitors, one may reduce corrosion and prevent sucker rod failure.

There are several types of corrosion agents:

1. Sour corrosion (H2S)

Hydrogen sulfide gas is very corrosive in nature. H2S will create deep corrosion pits (like hydrogen bombs) on the surface of the sucker rods, where minor damages such as dents, scratches, or rusted areas exist. H2S corrosion pits will cause sucker rods to break under loads. The black deposits on sucker rods are H2S bacterial buildup, which can literally destroy sucker rods and tubing strings in a short time.

Using H2S scavenger treatments may reduce H2S damages. High H2S conditions may be present in all fields that are subject to water flooding and/or holes in casing strings. Oversized rod guides will cause a choking effect and may create severe corrosion/erosion pits below and above the rod guides. Only one deep corrosion pit is enough to break the rods.

2. Sweat corrosion (CO2 corrosion)

Carbon dioxide appears like a worm and eats through the surface of wood. CO2 can create long pattern grooves, causing sucker rod damages.

3. Oxygen corrosion

Free air is introduced into wellbores during workover activities and fluid circulation. Free oxygen mixed with other elements will cause corrosion **(gas blowdown on surface)**

4. Surface electrical currents

Known as electrolysis, this is created by power lines and electrical storms.

Rods may become damaged and fail as the result of misuse. The enemies of steel sucker rods are too many. It is very difficult to keep operating a string of sucker rods for months without the high cost of maintenance. A pump speed of over ten strokes/minute on shallow wells with a small rod string may cause fatigue failure.

- Friction

Friction of any type should be avoided or minimized to prevent rod failure. Deviated holes are the major cause of friction failures. Use molded rod guides in deviated wellbores to reduce holes in tubing and sucker rod wear.

- Sand problems

Sucker rods are not suitable to perform in sandy wellbores. The well should be gravel-packed with a screen and liner to avoid rod stress and failure.

- Scale problems

Formation scale is created from reservoirs. Some wellbores cause alarming scale problems that literally damage rods, especially sucker rods.

- Salt problems

Some formation reservoirs have a tendency to produce saturated salt, oil, gas, and salt water. As the fluid cools off near the surface, the salt will drop off the solution and may pack off around the rods, causing the rods and pump to get stuck and become parted. Proper downhole treatment is recommended with the application of steamed water down the annulus to break down the salt composite.

- Sucker rod design

All downhole sucker designs must be carried out with the correct balance and by experienced and knowledgeable engineers to avoid rod string failure. The pump speed and SPM should be included in the rod design. The tapered rod design must be carried out with balance. The rod design may be tapered or straight and of one size.

- Paraffin problems

Paraffin is a greasy substance that comes from oil and gas products. As the fluid flows upward to the surface, the fluid will cool off, causing paraffin to come out of the solution, deposit around the rods, and block off the flow. Some wells will have considerable amounts of paraffin than others. Paraffin may pack off and cause parted rods.

High-content gas wells tend to deposit more paraffin because of the cooling action at the surface. Hot oil and hot steam injections down the annulus will remove and prevent paraffin buildup.

- Hammering blows on sucker rods

Hammering the rod boxes and tubing string is an ongoing bad practice in oil and gas operations without understanding the consequences (it is plain ignorance to break the rods or back off the tubing coupling). Never beat a rod box with a hammer. It will deform the rod box permanently and damage the threads. Never reuse a box that is hammered on (it is permanently deformed).

- Bending and buckling

Rods must be kept straight before going into a well. All sucker rods with kinks and bends must be discarded to avoid rod breaks because of load stress.

- Bad makeup torques

All sucker rods must be made up according to the manufacturer's makeup torque (use displacement cards provided by the rod manufacturer). Using a hand wrench is the proper way to make up the sucker rods before power tongue application.

- Thread galling failure

To prevent thread and sucker rod failure, the following steps must be carried out:

a) Inspect the pin thread and box thread to make sure they are clean and undamaged.
b) Never drag the rods on the ground.

c) Never walk on the sucker rod string without proper protection.

d) Do not allow the rods to hit or slap against the pipe racks when lifting.

e) The pin and rod protectors must be pulled off on the rig floor before making up the rods.

f) The rods must be kept clean without mud, sand, dirt, or rust. All the used rod couplings and rod pins should be cleaned with kerosene.

g) The rods must be properly lubricated before makeup (do not use a tubing compound on the sucker rod threads).

h) A 50/50 mixture of 40 wt oil and corrosion inhibitor will serve as a good rod lubricator (use the rod lubricant from the rod and pump supplier).

i) The rod threads must be kept very clean and lubricated using the rod lubricant before making up the rods.

j) The hydraulic power tongs must be rigged up properly to avoid bending and twisting the rods.

k) The rods must be stabbed properly—straight and without bending, bowing, and/ or leaning—before being made up.

l) The rods must be made up using a hand wrench first before using hydraulic power tongs to finish the final makeup torque (to avoid galling).

m) Galled threads will not make the makeup process easy and will make the box hot.

n) Avoid running or tripling out the sucker rods during high winds.

o) The makeup displacement

    cards must be used for the appropriately sized sucker rod.

p) Avoid over-torqueing or under-torqueing when making up the rods.

- **Gas locking (un-wanted downhole problem)**

Gas locks may occur in gassy wells as the result of poor well completion and tubing and rod landing design. When too much gas enters the pump chamber, the rods will beat against one another at the bottom, compressing the gas against the valves. Gas locking may reduce daily production. You will see high production one day and low production another day. It may break the rods, create holes in the tubing, and damage the surface and subsurface equipment. Gas locking can be avoided with proper downhole equipment designs using longer mud anchor joints (24' deep tube). Running gas anchor below the open perforations serves the best if possible

- Misapplication and excess pulling and stretching on the rod string

Often stuck sucker rods and subsurface pumps may create a condition that is not suitable for sucker rods. Excessive rod pulling, jarring, and twisting as well as fishing to release the stuck fluid pump may permanently damage the sucker rods. Did you read the section on safe sucker rod pulling?

- Fatigue, strain/stress, and overloading

One of the major causes of fatigue failure on sucker rods is the pumping speed and pick polished rod load. Fast reciprocating rods will have high fatigue stress failure and shorter useful lives. Do not overload the rods and the gear box. Rods should be designed with long and slow strokes per minute. (I prefer six to eight strokes per minute with the longest stroke length possible.)

Why a Polished Rod Breaks

- The system experiences corrosion attacks.
- The stuffing box pack-off is too tight (causing metal-to-metal friction).
- The carrier bar is not leveled, causing the polished rod to lean off-center.
- The wellhead is crooked, pulling the polished rod to one side.
- The polished rod clamp is too tight (they used a cheater pipe, causing cracks).
- Hammering blows are delivered at or near the break point.
- The pumping unit horse head is off the center of the wellbore.
- The polished rod is not at the center of the V-groove in the horse head.
- There is too much heat at the stuffing box (too hot for the polished rod).
- There are manufacturing defects in the casting.
- It is not a spray metal polished rod box (the polished rod box has more threads).

The Dynamic Load on the Sucker Rod String

The dynamic load is the force working against rod performance, including but not limited to the following:

- Hydrostatic column of fluid in the tubing string above the pump
- Weight of rods in the air
- Plunger friction
- Pack-off friction
- Rod friction
- Friction of solid buildup in tubing and around the rods

The reduction of pick polished load will allow one to lower the fluid pump deeper, if necessary, to increase production. Sucker rods encounter enormous problems from the shipping point to oil well operations. One of the major problems with sucker rod performance is the people who handle the equipment and run the operation in the oil field (lack of training and other things combined).

Continuous Steel Sucker Rods

In the early years, sucker rods were manufactured as continuous steel sucker rod strings. Because of many mechanical problems, these continuous sucker rods were eliminated. Later on, the rods were manufactured to be 25' long, with a threaded pin on one end and a coupling box on the other end of the rod (the coupling was actually made as part of the steel rod and could not be replaced). These rods were active in oil wells until 1998 (I was running them in the fields of Goose Creek in Baytown and the Thompsons fields of Texas).

The main problem with the coupling being part of the rod was that when the coupling wore down because of frication, it could not be replaced (the whole sucker rod had to be junked). "Modern" sucker rods are designed with pins on both ends. Sucker rods are delivered to a well location, with the sucker rods and couplings separately delivered. When the rod box goes bad, the coupling will be backed off and replaced with a new coupling accordingly.

When the rods break, the boxes should be backed off and reused.

Sucker Rod Dimensions

- Oil field sucker rod length: **25' to 30'**
- Oil field sucker rod size: **⅝", ¾", ⅞", 1", 1 ⅛", 1 ¼"**
- Oil field pony sucker rods: **2', 4', 6', 8', 10'**

Do not run oversized rod couplings in small-diameter tubing (need enough fishing room).

	Full-Size Box OD	Slim-Hole Coupling OD	Length
⅝" = 0.625"	1 ½"	1 ¼"	4"
¾" = 0.750"	1 ⅝"	1 ½"	4"
⅞" = 0.875"	1 13/16"	1 ⅝"	4"
1" = 1.00"	2 3/16"	2"	4"
1 ⅛" = 1.125"	2 ⅜"	2 ¼"	4 ½"

Production Tubing Sizes

Tubing Size	Tubing Weight	Inside Diameter	Capacity (bbl/ft)
1.900"	2.90#	1.610"	.0025
2.063"	3.25#	1.751"	.0030
2.375"	4.70#	1.995"	.0039
2.875"	6.40#	2.441"	.0058
3.500"	9.30#	2.992"	.0087
3.500"	10.20#	2.991"	.0083
4.500"	12.75#	3.958"	.0152

Sucker Rod Break Points

Sucker rods break for several reasons:

a) Corrosion pits (make the rods weak under loads)
b) Stress and fatigue caused by fast reciprocation or strokes per minute
c) Sudden change of load (water/oil ratio, sand and solids)
d) Defects of many kinds for many reasons

The most popular rod breaking points are the following:

a) Body break
b) Pin thread break
c) Sucker rod box (coupling) break

d) Wrench flat break (below the shoulder)

To fish and recover the broken sucker rod

out of the hole, you must run the correct-sized overshot or special tools to fish the rods out.

⅝" Parted Rod Fishing

⅝" body break   .625" grapple
⅝" pin break   1 ⅜" grapple
⅝" box   1 ½" grapple
⅝" slimehole pin or box   special tools

¾" Parted Rod Fishing

¾" body break   .75" grapple tool
¾" slimehole box   1 ½" = 1.50" or 1 9/16" = 1.5625" grapple tool
¾" full-size box   1.625" = 1 ¾" or 1 15/16" = 1.9375" grapple tool

⅞" Steel Sucker Rod Fishing

⅞" slick body break   .875" grapple
⅞" slimehole rod box   1 ⅝" = 1.625" grapple

KHOSROW M. HADIPOUR

⅞″ full-size rod box	1 13/16″ = 1.8125″ grapple
⅞″ reg. box	may not be fished inside 2 ⅜″ tubing (tight for overshot)
⅞″ pin break	need to fish for thread base shoulder (1 3/16″ = 1.1875″ grapple)

1″ Steel Sucker Rod Fishing

2 ⅞″ tubing = 2.221″ ID and 3 ½″ tubing = 2.992″ ID

1″ body break	1″ grapple
1″ slimehole box	2″ grapple
1″ full-size rod box	2 13/16″ = 2.8125″ grapple

The applications of fishing tools to recover steel sucker rods are as follows.

Fishing Stuck Pump and Sucker Rod String

Before attempting any downhole rods or tubing fishing, make sure the rig equipment is in good, safe operating conditions. (Check the cable and blocks to make sure they are safe for fishing rods and tubing.) The rig indicator must function with accurate readings. To prevent major accidents. zero the weight indicator prior to pulling upstream tension onto the fish.

The rod string is connected to the subsurface fluid pump under the dynamic load. If the pump becomes stuck for various reasons, it may cause the rods to get parted. Slow upward lifting without jarring will be more effective to reach better results. Do not jar while you are working the rods to pull the fish free. Jarring and pulling at the same time will exceed the upstream tensile force and may exceed the strength of the sucker rods. Over-pulling against formation sand and solids will squeeze out the liquid and will make it more difficult to pull the pump free.

Safe Pulling on Tapered Rod String

Always consider using a small rod string (¾", ⅞", 1" tapered rod string).

Safe pull (SPL) = cross-section area of the small rod × minimum yield value × safety factor + the weight of the rod string above the smallest rod section

There are several scenarios of fishing for stuck or parted sucker rods:

- Stuck pump and sucker rods
- Parted sucker rods
- Stuck pump and rods caused by parted tubing string
- Dropping rods in the tubing string
- Dropping rods in the casing string
- Stuck tubing, pump, and sucker rod caused by stuck TAC
- Stuck rods caused by broken rod guides
- Stuck rods caused by salt and paraffin buildup

We have the technology and know-how to resolve the worst fishing scenarios. I will briefly explain each of the above conditions.

- **Sucker rod stripping operation:**

If sucker rod string is not free falling freely it is due to the following problems:

a) Parted rods shallow (parted rods is due to fatigue, sand, paraffin etc.)
b) Stuck pump (sand, rod guides, and paraffin build up)
c) Parted tubing- parted tubing without tubing anchor may cause tubing to fall and cause sharp kinks on tubing (pulling rods often difficult without stripping the rods and tubing. Stripping the rod string and tubing is often unsafe, and difficult work (Backing off and cutting sucker rods and tubing under well pressure is possible)

Make sure you provide proper tools and equipment and safe practice during striping operation to shut the wellbore in case of flow emergency (stripping under well pressure may cause fire and cause injury)

Fiberglass Rods (Fiber Rods)

Fiberglass sucker rods are

- light,
- stiff,
- strong,
- straight,
- flexible, and
- noncorrosive.

The fiber rod is manufactured with a steel pin on one end and a steel coupling on the other end (similar to steel sucker rods). The ends of all manufactured fiber rods are made up with steel connections. Fiber rods are manufactured to be ¾", ⅞", 1", and 1 ¼", with a length of 37.50' (11.43 m).

Fiberglass Rod Data

Rod Size	¾"	⅞"	1"	1 ¼"
Rod Body	.738"	.858"	.980"	1.228"
Weight/Foot	.49 lbs.	.63 lbs.	.82 lbs.	1.4 lbs.
End Fittings	1.50"	1.50"	1.625"	2.0"

The technology and engineering of fiberglass are constantly changing. For proper installation and maximum loads, consult with the fiberglass manufacturer and local suppliers.

Sucker rods are the principal tool to transfer the reciprocating force from the subsurface fluid pump to the surface and from the surface to the subsurface equipment. Sucker rods are screwed together with sucker rod couplings. The rods are connected to the subsurface downhole pump and extended up the hole to the polished rod and up to the carrier bar at the horse head of the pumping unit above the ground.

Fiber rods are strong and resist against corrosion very well. They are light and will reduce the pick polished rod load because of their light weight. Fiberglass rods are designed in lengths of 37.50' (fiber pony rods are available).

Advantages of Fiberglass Sucker Rods

a) Lightweight compared with steel sucker rods
b) Resist against corrosion very well
c) Noncorrosive fiber body
d) Strong and tough
e) Straight

f) Flexible
g) Pin and couplings are made from steel
h) light to handle
i) Can be used in conjunction with steel rods (tapered string)
j) Pony rods available for spacing

Disadvantages of Fiberglass Rods

a) Must be put in tension while in service
b) Cannot bump against and/or pound on the rods
c) Will become parted, similar to steel rods
d) may become unglued from steel connection
e) Will split apart
f) More difficult to fish than steel rods
g) Must be kept away from sunshine. must keep in special containers
h) Skin must be protected from rods fiber hairs
i) Difficult to discard fiber junk rods (no one wants old fiber rods)
j) Environmental junk

III

CHAPTER

ARTIFICIAL FLUID LIFT TECHNOLOGY USING GAS LIFT EQUIPMENT

Artificial gas lifting is a very dependable, efficient, and environmentally friendly operation method. Highly compressed dry natural gas is the primary energy source of the artificial fluid lifting method. With the high quality and quantity of dry natural gas pressure, the gas lifting method becomes one of the most excellent and favorable artificial fluid lifting techniques.

Sour natural gas (H2S) may be utilized in the gas lifting process with full knowledge and understanding of corrosion and high safety risks. The gas lift method is operated using downhole gas lift valves. The gas lift valves are designed based on the following information:

- Operating depth
- Anticipated maximum daily fluid production
- Reservoir bottom hole pressure
- Static bottom hole pressure
- Flowing bottom hole pressure
- Reservoir PI(productivity index)
- Reservoir bottom hole temperature
- Volume of fluid subject to lift
- Density of fluid subject to lift
- Static fluid level

KHOSROW M. HADIPOUR

- Available gas lifting gas pressure (using gas compressor)

- Available gas lifting gas volume (using compressor)
- Production packer depth
- Tubing string size
- Casing string size
- Perforation depth
- Other helpful downhole mechanical lifting information

The tubing string is run in conjunction with an isolation packer and spaced-out gas lift mandrels and valves

to predetermined gas lift–designed depths as accurately as possible. The gas lift valves are normally pre-charged with N2 pressure and designed based on the available dry natural gas lift pressure, volume, and draw-down well information. The production packer is run at the bottom of the tubing string to isolate the gas lift equipment above the open perforations to avoid losing high-pressure injection gas lift gas into the open perforations and also to avoid back pressure on the formation. The gas lift valves will be spaced out and run on the tubing string (various sizes) based on the pressure rating and lifting depth above the open perforation.

Each gas lift mandrel/valve

is designed to be run and set at a specific operating depth on the tubing string and will open and/or close on a pre-calculated gas lift pressure (any mistakes in running a valve at the wrong depth will pose lifting problems). An effective gas lift method may be achieved with a good-condition casing string at a depth of 4,000'–12,000' (the casing string must be tested for leaks). Based on reservoir deliverability, a good gas lift design can lift fifty thousand barrels of fluid per day (do not attempt to gas-lift in any casing string with hole/s).

There are two types of gas lift valves/mandrels:

- Side pocket gas lift valves and mandrels

- Conventional gas lift valves and mandrels

Side pocket gas lift mandrels and valves are run in larger casings and tubing strings (the mandrels are larger and need more space). Side pocket mandrels are run on the production tubing to the required depth based on the gas lift design (side pocket valves can be installed using a slick wireline unit). As the name implies, side pocket gas lift valves are run and set through the tubing string and operate inside a side pocket of gas lift mandrels.

The side pocket mandrels that holding gas lift valves are designed so that the side pocket is positioned off the center of tubing string to avoid fluid cut and formation solids. The gas lift valves can be installed in the side pocket mandrels in the shop and/or using wire tools to run or remove the valve/valves using wireline kick-over tools.

Conventional gas lift mandrels are screwed and run on the tubing string, similar to side pocket gas lift mandrels. The valves are run and set and operate on the outside of the tubing string (gas lift valves are located in the annulus between the casing and the production string, considering the size of the casing string or annulus space between the tubing and the casing diameter).

The topmost gas lift valve is called the "kickoff valve," and the last gas lift valve in the well is called the "operating valve." A production packer (isolation tool) will be run below all the gas lift valves and at the end of the tubing string at a predetermined depth above the open perforation. The packer may be spaced with several joints of tubing above the isolation packer based on the gas lift design. A mechanical right-hand-set and straight pull packer will serve the best purpose in gas lift operations (this will avoid running the mechanical set and mechanical release packers). See the example and landing sheet on gas lift equipment.

The purpose of the isolation packer in gas lift operations is twofold:

- Avoid injecting gas lift gas into the open perforations
- Avoid high gas back pressure against the wellbore fluid

Gas Lift Equipment Running Procedure

- Hold and document safety meetings.
- Check the safety anchors on location. Test the anchors at 30,000 lbs.
- Use a base beam if there are no permanent anchors on location.

- Move in and rig up the workover rig. Secure the rig to safety anchors.
- Spot the rig pump and tank. Fill up the rig tank with clean produce water.
- Circulate and kill the well if necessary with clean fluid.
- Nipple down the wellhead. Nipple up the BOP; test at 200# (low) and 3,000# (high).
- Never shortcut on testing safety equipment.
- Pull out any production equipment left in the wellbore.
- Make up and trip in the hole with a bit and scraper

  on the tubing string. Test and drift the tubing string as required.
- Clear and drift the casing and open perforations. Leave at least 100′ of the rat hole below the open perforations.
- Circulate the wellbore with clean field produce water or 2% KCl water.
- Pooh with a tubing string; lay down the bit and scraper assembly.
- Prepare to test the casing string for hole/s (use casing MIT test if necessary).
- Prepare to run the gas lift equipment as required based on the gas lift design sheet.

The gas lift equipment (from the bottom up) consists of the following:

- o Short mule shoe

- o Isolation packer (right-hand-set and straight pull packer preferred)
- o Seating nipple or seating shoe (optional)
- o Joints of tubing based on the gas lift running procedure sheet
- o Operating valve
- o Joints of tubing with spaced-out gas lift valves up the hole
- o Kickoff valve on the very top as designed
- o Tubing joints to the surface landing joint
- o Wellhead equipment

- Strap the tubing string, isolation packer, and gas lift valves as accurately as possible (measure and drift anything that goes in the well).
- Make up and trip in the hole with the isolation packer, seating nipple, tubing, and gas lift mandrels according to the recommended gas lift design sheet.
- The operating valve will be at the bottom and the kickoff valve on top of the tubing string. The rest of the valves will be spaced out in between according to pressure ratings on the gas lift design. As many as six to sixteen gas lift valves and mandrels may be run in some oil wells to successfully lift the wellbore fluid.

Gas Lift Equipment Running Procedure:

- Prepare to test and run the tubing string and gas lift valves in the well.
- Apply a light pipe thread compound to the pin ends only.
- Make up the tubing and gas lift mandrels by hand first; then use hydraulic tubing tongs to achieve the proper makeup torque as recommended.

- Make sure all the gas lift valves are installed properly and at the correct measuring depth based on the tubing tally and the gas lift design sheet. All the mandrels with the correct pressure rating valves are marked and numbered to avoid running mistakes. (Do not be confused by the valve numbers.) On the design sheet, Valve #1 (the operating valve) is the first valve spaced above the packer. On the tubing landing report, Valve #1 is the kickoff valve at the surface. If there is any mistake in running any of the valves in the hole, the equipment will not function properly, and you may have to pull the well again to check and correct these mistakes.
- Run the valves according to the gas lift spaced design sheet (normally, a gas lift rep will be present when running the equipment in the well to avoid mistakes).
- Test the tubing string and gas lift equipment going in the well as required.
- Avoid damages to the gas lift mandrels and/or gas lift valves while making up the mandrels on the tubing string.
- Finish tripping in the hole with the packer, tubing, and gas lift equipment to the desired operating depth (all the gas lift mandrels with gas lift valves are normally stamped on at the shop to avoid mistakes). Gas lift mandrels are normally set on the rig floor or on the ground in the order to be run in the well (first in and last out).
- Check and run all the mandrels according to the gas lift design sheet.
- Once the equipment is run in the well, space out and set the isolation packer (avoid too many right-hand turns to set a packer).
- Avoid jarring or beating on the packer to set and/or to isolate. You cannot test the packer to find out if the packer is holding pressure or not (too many holes are open on the string).
- Check the well for flow before removing the BOPs.
- Remove the BOPs and clear around the wellbore.
- Nipple up the wellhead equipment and install a high-pressure gas lift line and the production flow line to the wellhead (make sure the wellhead flange is free of leaks to avoid losing natural gas lift gas).

On gas lift well operations, there are two separate flow lines at the well: one coming to the well (gas lift line) and one leaving the well (production flow line from the well to the tank battery). A 2″ high-pressure gas supply line usually comes from a gas compressor to the well. A natural gas meter recorder and gas choke assembly are installed on the line before the gas is injected into the casing annulus. A larger production line is required from the well to deliver the production fluid to the tank battery. A larger flow line is preferred to reduce back pressure. I also prefer looping (arching) the line as soon as fluid passes through Christmas tree at the well

Two pin pressure recorder and gauges are installed to monitor pressure and fluid lifting information (one pin to the annulus and one pin on the tubing string on top of the Christmas tree). The production line is the flow line from the well to the production facility and must be a larger pipe with minimum restrictions to avoid back pressure against the reservoir and gas lift operations system (avoid too many Ls and Ts on the production line). Install a gas meter and choke with the related equipment to measure and inject high-pressure dry gas *slowly* into the casing/tubing annulus to displace the annulus liquid (this will prevent cutting the valve out).

Conventional gas lifting is one of the most efficient and versatile artificial fluid lifting methods in oil patches (based on my twelve years of continuous gas lift operating experience while working for the Gulf Oil and Chevron Company in Thomson, Texas).

The artificial gas lift method has many great advantages over other lifting methods:

a) Low installation and maintenance cost
b) Environmentally friendly and safe operation (clean surface area)
c) Low lifting costs compared with other artificial lifting methods

d) Lower surface and subsurface equipment problems
e) Easy to operate, troubleshoot, and monitor the well performance
f) Excellent ability to handle nonconsolidated sand and solids, better than all other types of artificial lifting systems (unbelievable)
g) Ability to reduce reservoir bottom hole pressure (looks like a natural flowing well)
h) Ability to lift high fluid volume as well as low production volume, ranging from fifty to fifty thousand barrels of fluid per day (based on the reservoir drawdown, fluid characteristics, and tubing string deliverability)
i) Tubing string fully open throughout
j) Ability to run tools through the tubing to check for or wash perforations using a wire line or coiled tubing (on mechanical artificial lift—such as beam pumping, jet pumping, submersible fluid lifting, and PCP pumping—you will not be able to check the perforations freely without pulling production equipment)

k) Ability for continuous lift as well as intermittent flow lifting design techniques

l) Best application for wells with high gas/oil ratios compared with other types of artificial lifting (you do not have to worry about gas surge and/or gas locking problems)

m) Highly effective and efficient lifting method in wells with dirty fluid and sandy wellbores

n) Can be used efficiently over a wide range of wellbore conditions, such as horizontal and deviated wellbores, with a high volume of fluid with sand production without the fear of mechanical wear or stuck equipment

Principal Reasons for Artificial Fluid Lifting

Most oil and gas wells may continue to flow naturally with the formation and/or reservoir gas pressure for some time. As the well produces high fluid volumes for quite some time, the formation pressure and reservoir energy begin to decline, causing the well to stop flowing. When the reservoir pressure declines more than the hydrostatic column of fluid in the wellbore, the well will stop flowing. In this case, the well fluid will stall at some point below the wellhead. Once the well stops flowing fluid up the surface, artificial lifting will be necessary to keep the well in production to recover oil and gas out of the reservoir.

When you are gas-lifting a well, the well seems to flow naturally; you don't have to worry about too many mechanical moving parts on the surface and/or subsurface, unlike with rod pumping or submergible pumping. Most problems such as rod wear, parted rods, tubing wear, stuck pumps, gas locking, water hammering, or high electric costs will be eliminated. The gas lift method is not only used to lift fluid in oil wells but can successfully be applied to water wells, higher water-producing gas wells, and/or backflow salt water disposal wells to remove sand and solids. Gas lift valves can effectively be used in downhole chemical treatments.

The gas lift method consists of the following:

A) Surface equipment system
B) Subsurface equipment system

The gas lift surface system consists of the following:

- Gas lift compressor

    and high-pressure natural dry gas availability
- Gas lift line from the compressor to the well
- Surface production equipment, such as the flow line and production separation equipment (separators, heat treater, and free water knockout gun barrels and stock tanks).

The gas lift subsurface system may consist of the following:

- Tubing string (2 ⅜″ to 4 ½″)
- Casing string (4″ to 13 ⅝″)
- Gas lift equipment (side pocket or outside gas lift valves and mandrels)

The gas lift method may be applied to lift fluid in various wellbores ranging from 2,000′ to 12,000′ deep using as few as two gas lift valves to fifteen or more gas lift valves and mandrels as needed (depends on the reservoir depth and formation drawdown characteristics).

How a Gas Lifting System Works

Gas lifting is a simple and efficient artificial fluid lifting system (it is not the same as mechanical artificial fluid lifting; there are no surface and/or subsurface moving components to worry about). Gas lifting operates by injecting high-pressure compressed dry natural gas at a constant low injection rate down the casing annulus through the gas lift valves and U-tubing fluid out of the production tubing string. The injected gas works according to the U-tubing principle from the annulus up to the tubing string. Gas lifting will reduce fluid density via the aeration of hydrocarbon fluid to the surface. Operating a well using gas lifting seems similar to doing so using normal flowing well operations.

How a Gas Lift Works

On the rig up configurations, normally, a two-inch gas line is run from a high-pressure compressor station and branches out to the well subject to artificial gas lifting. At the well, a gas choke and gas lift meter will be added and installed to monitor and control the volume of gas input into the well. You can calculate the volume of gas going into the annulus and the volume of gas coming out of the wellbore (the extra produced gas will return to the sale line and/or compressor station).

To effectively gas-lift a well, the borehole must be cleaned up first. Normally, the wellbore is completed and cleaned up using a drill bit and a casing scraper. The casing string must be tested to ensure it is in good standing above the open perforations. Testing the casing string is a necessary step to check the casing's MIT to avoid losing the gas into the casing leaks and to avoid sanding up the downhole equipment in the well. Never gas-lift old wells with casing hole/s, including squeezed holes. After completing and cleaning the wellbore to the plug back depth (below the open perforations), run the gas lift equipment into the well. Normally, the gas lift equipment is designed and made up in the gas lift shop and then delivered to the well location to be run in the well.

There are two types of gas lift valves and mandrels in oil field operations:

- Side pocket gas lift mandrels and valves (the valves are inserted inside the gas lift mandrels)
- Standard outside gas lift mandrels and valves (the valves are located outside of the mandrels, between the tubing and the casing annulus)

The outside gas lift valves are normally made up on the outside of the gas lift mandrels before they are delivered to the well location.

Side pocket mandrels and valves can be run in two ways:

a) Seat the valves in the side pocket and run the mandrels on the tubing string.
b) Run the side pocket mandrels on the tubing string and later run the valves on the wire line to seat the valve/s inside the side pocket mandrels. I prefer to seat the side pocket valves in the shop before running the mandrels into the well.

Normally, the gas lift valves are carefully numbered in order according to the landing depth and pressure rating to be run into the well (any mistake in running the valves to the incorrect depth will cause the well not to lift properly). Most often, a gas lift equipment technician may arrive with the gas lift equipment and stay on the well location to supervise while making up and running the valves in the hole according to the design sheet (he will charge you for service time). The valves are checked and made up on the rig floor to ensure that all the valves are put in the well correctly and according to the gas lift calculated design sheet.

The gas lift equipment may consist of the following:

- Tubing strings of various size and grades
- Gas lift mandrels
- Gas lift valves
- Production isolation packer

The mechanics of the gas lift operation can best be illustrated below. After the well is cleaned up with a bit and scraper, the gas lift equipment is run on the tubing string into the well as follows:

a) Measure and keep an accurate tally on the tubing string and gas lift equipment going into the well.
b) Test and rabbit anything going into the well.
c) Trip in the hole with the correct-sized mechanical production packer. Generally, mechanical right-hand-set compression-type and straight pull packers are run in the gas lifting wellbores

(the mechanical set with the straight pull-up packer is easy to set and release without too many rotations to avoid twisting off the tubing and damage the outside gas lift valves). The purpose of the isolation packer is to prevent fluid from entering the annulus, keep the gas from going down into the perforations, and avoid high back pressure against the reservoir fluid (the isolation packer must be reliable and must not leak fluid).

d) Space out before tripping in the hole with the gas lift equipment.

e) Check the well for flow. The well fluid must be static below the surface before starting in the hole.

f) Make up the production packer on the tubing and lower into the hole (check the rubber elements to ensure they are in good condition).

g) Lubricate and apply a light thread compound on the pin threads of the gas lift mandrels only.

h) Apply the thread compound properly on the tubing pins only.

i) Make up each gas lift mandrel with the valve on the tubing string by hand first; then apply hydraulic tongs to deliver the proper makeup torque.

j) Avoid damages to the valves. The short space between the valves and the tubing tool joint will be good enough to allow you to properly make up the mandrels. The gas lift mandrels and valves are spaced and numbered according to landing depth on the running sheet (make sure the valves are run as recommended).

k) Trip in the hole with the packer and gas lift equipment on the tubing string as recommended.

l) Run in the hole with the gas lift equipment slowly to avoid sudden stops. This may prevent accidents and damage to the valves. Once the gas lift equipment reaches the required landing depth, check the well for flow (the well fluid must be static during tripping).

m) Always remember that there is a significant difference between the well fluid flowing pattern and the pipe displacing pattern before a well kick occurs.

n) If the well fluid is static and no fluid flow occurs, then space out and set the packer.

o) Remove the BOPs and install the tubing head. After setting the packer, you may not be able to test it correctly down the annulus (all the submerged valves may be open because of the hydrostatic fluid). You may be able to test the packer using the circulating technique only.

p) Avoid setting the packer on a casing collar. The packer should be set across a cement bonding interval in the casing and off the casing collars.

q) Nipple down, nipple up, and install the tubing head. Make sure the tubing head bolts are tightly made up to avoid gas leaks out of the wellhead flanges.

r) Install the gas lift line and a choke assembly to the casing annulus string.

s) Install the flow line from the well to the tank battery. Avoid too many Ts and sharp bends on the flow line from the well to the tank battery to avoid back pressure (the application of an arched flow line above the wellhead is recommended, if possible).

t) Install a gas choke assembly and a gas lift meter on the gas line.

u) Install a fluid meter (a two-pin Burton-type meter on the production line or Christmas tree is needed to monitor fluid flow patterns). Do not use the back pressure valve and/or choke assembly on the gas lift system.

Normally, after well completion and/or workover, the tubing string and the gas lift equipment will be submerged into the wellbore, full of heavy fluid to surface. All the submerged valves below the fluid surface should be considered open from the top to the bottom because of the hydrostatic column of fluid above the valves.

Gas lifting wells may be designed using several gas lift valves based on certain gas lift requirements. Some wells may require as many as fifteen gas lift valves to operate the well (based on the well depth and reservoir drawdown characteristics). The gas lift valves above the operating valve are mostly for unloading the liquid in the annulus area. The functions of all the valves are very important to unload liquid from the annulus.

In gas lift operations, the top valve is call the kickoff valve, and the bottom valve is called the operating valve. (The last two bottom valves may alternate and are very important to lift wellbore fluid efficiently.) Since the annulus space between the tubing and casing and the tubing string is full of fluid, all the gas lift valves should be considered open because of the hydrostatic column of fluid above the valves.

The fluid must be displaced by forcing the dry gas down the annulus to push the wash water or wellbore fluid out of the annulus and the tubing string to reduce hydrostatic pressure and to close each of the shallow located valves. The sequence of how a valve operates is described below. To start with, the adjustable gas choke at the surface on the gas line is closed shut under high gas pressure, and no gas enters the annulus. Figure I shows the wellbore full of liquid that must be displaced with gas lift gas. All the valves are in the open position.

To displace fluid, we start with opening the gas lift choke valve at the surface and begin injecting high-pressure gas down the annulus very slowly and at a constant rate to avoid damaging or cutting the subsurface gas lift valves (have patience when you are displacing the annulus liquid with gas lift gas to avoid damage to the gas lift valves; slow and constant input gas will displace and unload the liquid more effectively).

All the gas lift valves are assumed to be open because of the hydrostatic column of fluid above the gas lift valves in the casing annulus and the tubing string. Figure I shows that the gas lift gas choke at the surface on the gas line is closed, and no gas enters the casing annulus. The fluid in the well is in a static condition just below the surface. All the gas lift valves below the surface are assumed to be open because of the hydrostatic column of fluid above the valves.

Advantages of Gas lift:

- Clean operation and friendly environmental sight
- Easy to manage and operate (similar to a flowing well)
- The best artificial system to handle formation sand and solids
- Good for lifting one hundred(100) barrel to thirty thousands (30000) barrel of fluid per day

- Easy access to wellbore for any well interventions
- Options for side pocket mandrels and/or conventional mandrels

The Disadvantages of Gas lifting System

➢ Require a Good standing Casing string (must not have a casing hole)
➢ Gas lift mandrels are subject to cut or washed at the injecting point
➢ Require a minimum 500 psig to 1500 psig dry natural gas

see pictures

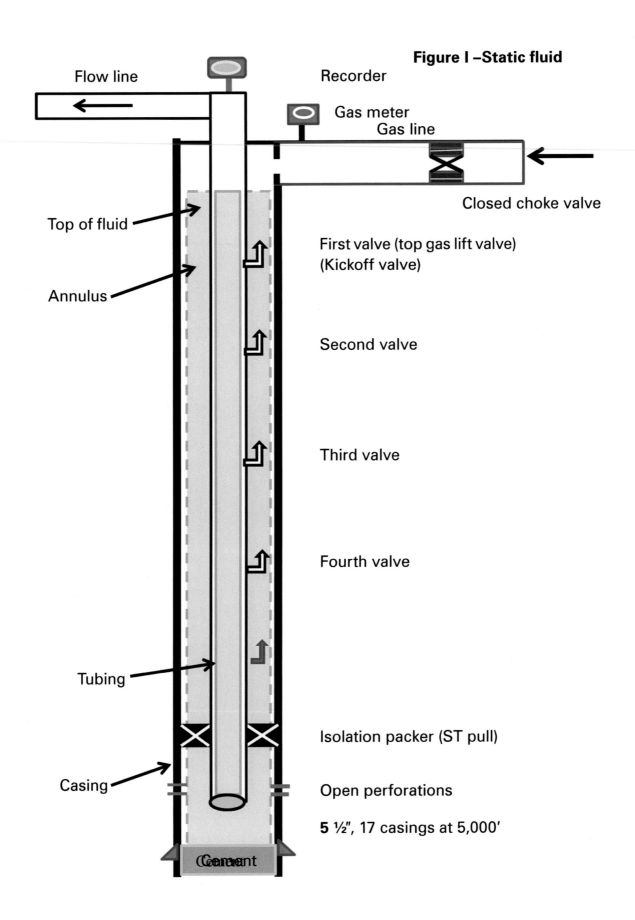

Figure I –Static fluid

Flow line

Recorder

Gas meter
Gas line

Closed choke valve

Top of fluid

First valve (top gas lift valve)
(Kickoff valve)

Annulus

Second valve

Third valve

Fourth valve

Tubing

Isolation packer (ST pull)

Casing

Open perforations

5 ½", 17 casings at 5,000'

Cement

Before You Start Unloading the Fluid Out of the Well Annulus . . .

a) Check all the surface equipment to avoid leaks during gas lift operations. Install new charts to calculate the input and output gas.
b) Fully open the master valve on the casing string (annulus).
c) Fully open all the valves on the Christmas tree and the valves on the tubing string (open the master valves and wing valve; do not cheat while opening the valves).
d) When injecting gas, you may calculate the volume of gas going down the annulus.
e) Check and adjust the gas lift gas going through the choke into the casing tubing annulus. When high-pressure gas forces the liquid down through the opened kickoff valve and valves below, the liquid is pushed through the valve into the tubing string and up to the surface.
f) The gas pressure will force liquid out of the open gas lift valves, especially the top valves and the U-tube, displacing the liquid from the annulus through the open valves and up the tubing string to the surface (the produced liquid may be wash water or dead wellbore fluid).

It is important that the annulus be unloaded as slowly as possible to keep the fluid from cutting the gas lift valves (a slow and constant gas injection rate is required). The liquid level is static at the surface inside the annulus and in the tubing string, with all the subsurface valves open because of the hydrostatic column of fluid above the valves (Figure I).

Gas begins forcing the fluid down the annulus and the U-tube out of the open gas lift valves and ports and out of the tubing string (Figure II). No formation fluid is expected at this point because of higher hydrostatic fluid pressure in the annulus and in the tubing string. Once the liquid is forced out through the open top gas lift valve (the kickoff valve) and into the tubing string, the gas will aerate the liquid, causing a significant decrease in fluid density and hydrostatic pressure. No formation liquid from the perforations is expected at this point.

As long as the hydrostatic column in the tubing and the annulus is greater than the reservoir pressure, the fluid will be static below the surface (no formation fluid). The reduction of pressure will help the open valves below the top valve unload the fluid more rapidly and efficiently (Figure II). Once the fluid level reaches the top gas lift valve (the kickoff valve), the gas-aerating action of the fluid will begin, causing a reduction in fluid density and the hydrostatic fluid column in the tubing string.

The gas lift gas will force and sweep off the water out of the casing annulus from the surface through the first and second valves and will continue displacing liquid until the liquid passes through the second and third open valves. When the fluid starts moving out of the second and to the third gas lift valve, the kickoff valve (top valve) will be forced close because of the pressure change. All the gas lift gas will continue down the annulus through the second and third valves to unload the liquid (Figure II).

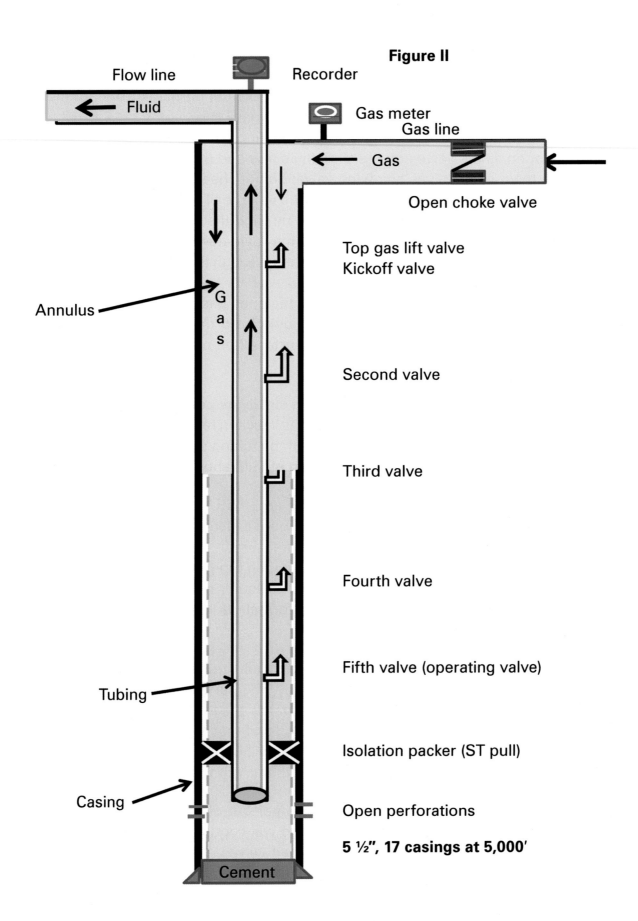

Figure II

Flow line   Recorder

← Fluid

Gas meter
Gas line

← Gas

Open choke valve

Top gas lift valve
Kickoff valve

Annulus

Gas

Second valve

Third valve

Fourth valve

Fifth valve (operating valve)

Tubing

Isolation packer (ST pull)

Casing

Open perforations

5 ½", 17 casings at 5,000'

Cement

All the gas lift gas will continue down the annulus to the second and third valves to unload the liquid (Figure II). As the gas lift gas forces the liquid through the second and third gas lift valves and into the tubing string, the hydrostatic column of fluid will be reduced inside the annulus and in the tubing string because the gas aerates the liquid while move up the tubing string. In this case, reservoir fluid pressure may start up the tubing string slowly.

Dead liquid passes through the second valve onto the third gas lift valve and into the tubing string. The annulus becomes full of dry gas from the surface to the third valve, causing the second valve to close shut because of the change in pressure (Figure III). The gas lift gas pressure will continue forcing the liquid down through the third and fourth open valves. Once the liquid passes the third and fourth gas lift valves, the third valve will be closed, and all the gas will be directed down to the fourth and fifth gas lift valves (operating valve). In some cases, the formation fluid may start moving in with oil and gas at the surface. The fourth valve may act as the operating valve until the dry gas reaches the last operating (fifth) valve (Figure IV). During the unloading operation, you may open the needle valve at the surface to see and feel the gas lift fluid flow at the surface. You may adjust the gas input as required.

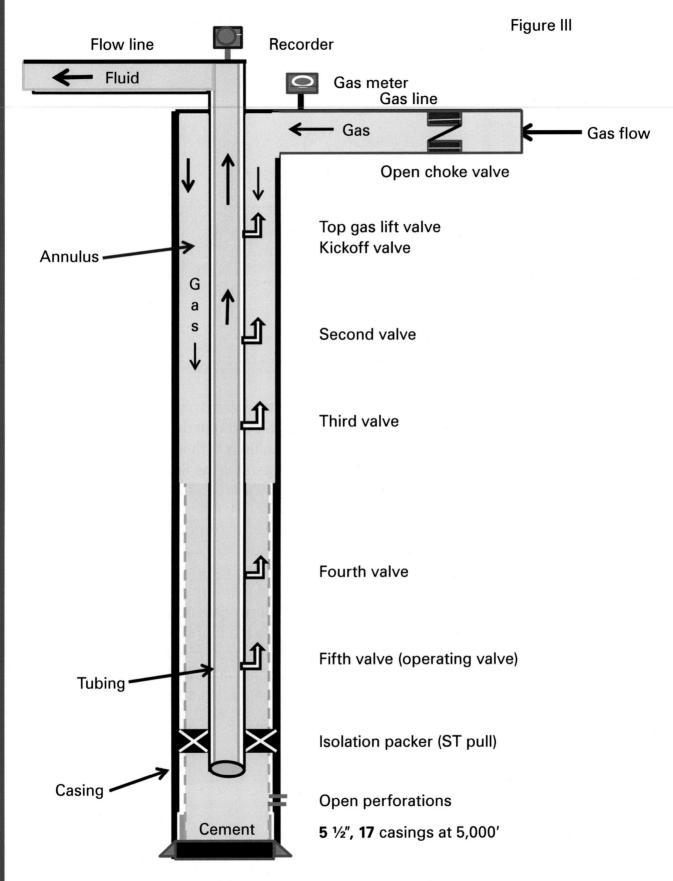

Figure III

Figure IV

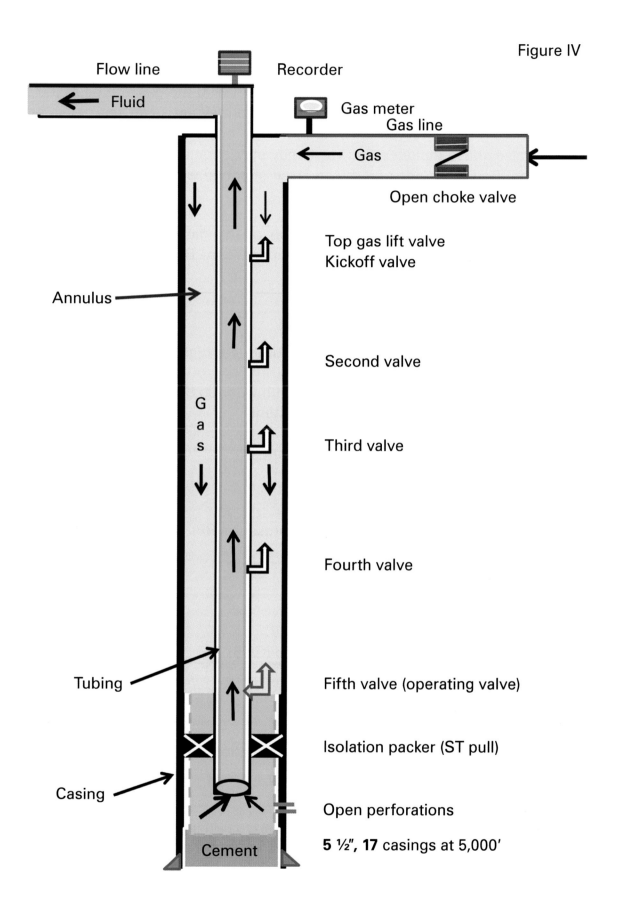

Flow line

Recorder

Fluid

Gas meter

Gas line

Gas

Open choke valve

Top gas lift valve
Kickoff valve

Annulus

Second valve

Gas

Third valve

Fourth valve

Tubing

Fifth valve (operating valve)

Isolation packer (ST pull)

Casing

Open perforations

5 ½", 17 casings at 5,000'

Cement

Once the liquid passes the third and fourth gas lift valves, the third valve will be closed, and all the gas will be directed down to the fourth and fifth gas lift valves (operating valve). In this case, the formation fluid may start moving in with the gas and oil at the surface. In some cases, the valve above the operating valve may act as an alternative operating valve until the dry gas is forced down to reach the operating valve (Figure IV). In some cases, the valve above the operating valve may stay open and must be forced closed by pinching. The gas lift gas must be dry and free of condensate liquid to avoid cutting the valves (gas lift, not liquid lift).

To better understand the principal operation of the gas lifting system, some basic knowledge of reservoirs and formulas may be necessary. Basically, there are three types of oil- and gas-producing gas reservoirs:

A) Water drive reservoir

As the name implies, the water drive reservoir produces higher salt water with oil and gas.

- As oil is produced, it will be replaced with water in the reservoir.
- The reservoir pressure stays fairly constant.
- The PI stays constant.
- The gas/oil ratio is fairly low and constant.

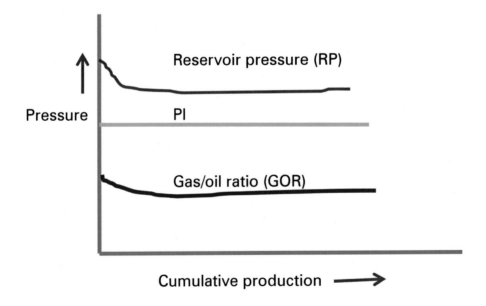

B) Depletion drive reservoir

The depletion drive and/or solution gas drive reservoir will have several characteristics:

- Constant volume with no water encroachment (when the reservoir pressure drops below the bubble point, it will cause two phase flow characteristics, producing oil and free gas)
- Decline in PI as the production declines

- Increase in reservoir gas as the oil production declines

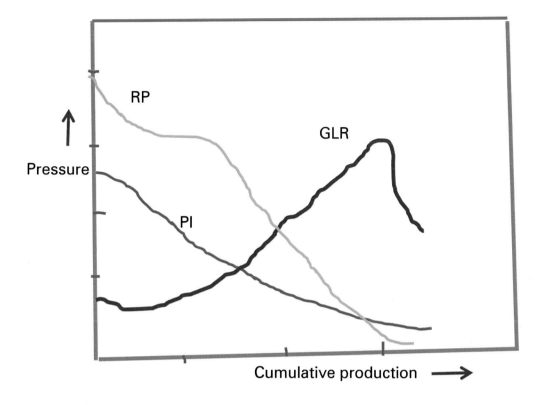

C) Gas-cap expansion drive reservoir

The gas cap will expand as the well production declines. The gas/oil ratio will be fairly constant at the beginning. The gas increases as the oil production declines.

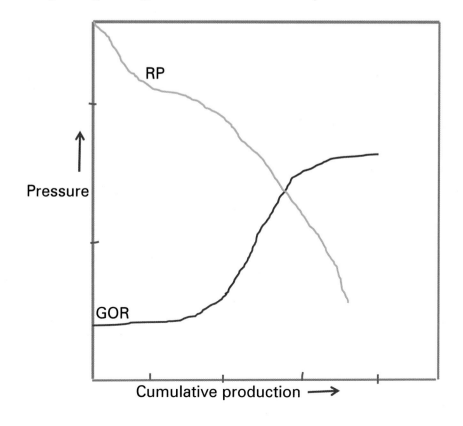

The basic formulas used in gas lift designs are as follows:

a) Hydrostatic pressure (0.052 × depth × fluid density)
b) Productivity index

$$PI = \frac{Q}{SBHP - FBHP} \qquad SBHP - FBHP = Drawdown$$

PI = productivity index (production rate (BLPD/PSI))

Q = production rate (total production per day)

SBHP = static bottom hole pressure (pounds per square inch)

c) $SBHP = FBHP + \dfrac{Q}{PI}$

FBHP = flowing bottom hole pressure (pounds per square inch)

d) $FBHP = SIBHP - \dfrac{Q}{PI}$

$$PI = \frac{total\ production\ rate}{drawdown}$$

Example

Production rate = 400 barrels/day

FBHP = 2,000 psi
SBHP = 3,500 psi

$PI = \dfrac{400}{3,000 - 2,000} = 0.4$ barrels per day/psi

If you are unable to obtain the PI, you may use the following formula with the two flow test rates and FBHP:

$$PI = \frac{Q1 - Q2}{FBHP\ 2 - FBHP\ 1}$$

First test rate: Q1 = 350 BLPD and FBHP 1 = 1,800

Second test rate: Q2 = 400 BLPD and FBHP 2 = 1,550

$$PI = \frac{Q1 - Q2}{FBHP\ 2 - FBHP\ 1}$$

$$PI = \frac{350 - 400\ BLPD}{1,550 - 1,800}$$

$$PI = \frac{-50}{-250}$$

PI = 0.20 BLPD/psi

On fluid hydrostatic pressure, the hydrostatic pressure is the weight of the column of fluid at a specific depth. Hydrostatic pressure does not depend on width; it depends on height.

HP = .052 × density × height

One gallon of fresh water is equal to one pound, one foot of height, or 231 cubic inches.

Example

What is the hydrostatic pressure at 6,000' TVD of the well with a fluid density of 11.2 lbs. per gallon?

HP = .052 × 6,000' (TVD) × 11.2 = 3,494.4 psi

What is the pressure gradient? The fluid gradient is the measurement of pressure for a given vertical fluid height. The density of fresh water is measured to be 8.33 gallons per foot.

Static fluid gradient = .052 psi/ft × fluid weight (fluid density) lbs/gallon

Fluid gradient = .052 × 8.33 = 0.433 lbs/gallon

All the fluid gradients ranging from 0.433 to 0.450 are considered as normal gradients. High pressure gradients are considered as abnormal pressure gradients. Fresh water weighs 8.33 lbs. per gallon, and salt water starts with a weight of 8.40 to 9.20 lbs. per gallon.

Static pressure gradient (psi/ft) = Specific gravity × 0.433 (psi/ft)

Hydrostatic pressure = Static fluid gradient × depth

Hydrostatic pressure = Gs (psi/ft) × h (ft)

Example

What is the HP at 6,000' of fluid with a pressure gradient of 0.433 psi/ft?

Hydrostatic pressure = .433 × 6,000' = 2,598 psi

If the fluid density is in lbs/ft³ . . .

Pressure gradient (psi/ft) = fluid density (lbs/ft³) × 0.006944

Example

If the fluid density is 120 lbs/ft³ . . .

Pressure gradient (psi/ft) = 120 × 0.006944

Pressure gradient = 0.833 psi/ft

or

Pressure gradient = $\frac{120 \text{ lbs/ft}^3}{144}$ = 0.833 psi

Pressure gradient (psi/ft) = fluid density $\frac{0.052 \text{ psi/ft}}{\text{lbs/gal}}$

Example

If the fluid density = 9 ppg, then the pressure gradient (psi/ft) = 9 × 0.052 = 0.468.

$0.052 = \frac{12}{231} = .051948$

The fluid-specific gravity is the ratio of the weight of the fluid to the weight of fresh water.

Pressure gradient= specific gravity of fluid x fresh water gradient 0.433(psi/ft)

Specific gravity (ppg)= fluid weight divided by 0.433!

Example

If the well fluid has a specific gravity of 1.12 and that of fresh water is 0.433 . . .

Gs (psi/ft) = 1.12 × 0.433 psi/ft = .485 psi/ft

The static gradient of downhole fluid properties can be calculated based on the following:

- Specific pressure gradient
- Fluid density (fluid weight)
- API gravity

The value of fluid gradient is very important in gas lift calculations. Each of the above known values can be converted to fluid gradient. Fluid density is the weight of a fluid for given volume in (lbs/gallon)

<u>Example</u>

If the well fluid is 12 lbs/gal (ppg) . . .

Gs (psi/ft) = 12 ppg $\times$ .052 $\dfrac{\text{psi/ft}}{\text{lbs/gal}}$ = 0.624 psi/ft

Pressure at 8,000′ = 8,000 $\times$ .624 = 4,992 psi

Pressure at 8,000′ = .052 $\times$ 12 $\times$ 8,000 = 4,992 psi

The pressure gradient can be calculated from the API gravity, which is expressed in degrees and is not related to fluid viscosity.

Fresh water: 8.33 with Gs = .433 psi/ft = API 10′

$$Gs = \frac{(141.5)}{(131.5 + API)} \times 0.433 \text{ (fresh water)}$$

What is the pressure of 2,000′ of crude oil with an API gradient of 35 degrees?

$$Gs = \frac{(141.5)}{(131.5 + 35′)} \times 0.433 \text{ psi/ft}$$

$$Gs = \frac{141.5}{166.5} \times .433 = 0.368 \text{ psi/ft}$$

0.368 $\times$ 2,000′ = 736 psi

Wellbore Gas Pressure and Temperature

Gas measurements can be expressed as standard cubic feet (SCF), which refers to the volume of gas occupied at standard conditions in atmospheric pressure at 60°F.

MCF = 1,000 SCF of gas

MCF = 1,000 × SCF

$$\text{FGOR (formation gas ratio)} = \frac{\text{Total formation gas (mcf/day)}}{\text{Oil volume (barrel)}}$$

$$\text{GLR (gas/liquid ratio)} = \frac{\text{Total gas volume (scf/day)}}{\text{Total fluid volume (bbl/day)}}$$

$$\text{GOR (SCF/BBL)} = \frac{\text{Gas volume SCF}}{\text{Total volume of oil (Q)}}$$

$$\text{Geothermal gradient temperature (100°F)} = \frac{\text{BHT (°F)} - \text{SST (°F)} \times 100}{\text{Depth}}$$

BHT = bottom hole temperature across perforations

SST = static surface temperature

$$\text{BHT} = \text{SST (°F)} + \frac{\text{Geothermal gradient (°F/100)} \times \text{perforation depth}}{100}$$

Well depth = 9,000'

SST = 90°F

Geo gradient = 1.4°F/100' (different at various spots)

$$\text{BHT} = 90°F + \frac{(1.4°F/100') \times 9,000'}{100} = 216°F$$

Review: How the Gas Lift Valves Operate

In gas lift operations, there is a "kickoff valve" at the surface (the first valve on the top), and there is an operating valve at the bottom above the packer. Several other valves are spaced out in between these two valves. To start with, the kickoff valve will open at the designated pressure and displace fluid out of the casing/annulus and kickoff valve into the production tubing string. When the liquid is displaced via gas lift pressure, the valves at the surface will be closed shut. When dry gas reaches the last or deepest valve, the operating valve, the opened operating valve will begin lifting the well fluid. If there are a few joints below the operating valve (between the packer and the operating valve), the fluid behind the fluid above all the blank joints will remain static above the packer for the duration of the gas lifting operation.

The casing annulus liquid will be displaced by clean, dry, natural gas through the annulus and all the closed shut valves because of the change of pressure from the surface gas lift kickoff valve to the operating valve. The operating valve is the last valve wherein the lifting actions begin by taking the injected gas and displacing the wellbore fluid to the surface.

The gas lift operation will continue by injecting constant dry gas down the casing, through the operating valve, and up the tubing string, displacing the well fluid of oil, water, and natural gas to the surface. The gas lift gas must be free of condensate or any injected liquid for an effective and efficient lifting operation (the valves are designed based on dry gas without liquid).

The gas lift method is capable of producing a large volume of fluid (Thirty thousand barrels) with or without formation sand and solids. This method will displace sand and solids better than any known artificial lifting method based on my twelve years of practical gas lifting experience.

ARTIFICIAL LIFT TECHNOLOGY USING JET PUMP AND HYDRAULIC PUMP

Hydraulic jet pump is an innovative artificial lifting method to extract oil, gas, and water out of the wellbore with good efficiency. The principal operating mechanics of jet pumping is to pump clean fluid (water) under high pressure down the tubing string and through a jet nozzle to create the "Venturi effect" and high differential pressure to lift fluid, circulating oil, gas, and water up the annulus and into the surface production facility. When a high-pressure fluid is pumped from the surface through a narrow jet nozzle at the subsurface pump, it will create the Venturi effect, causing high differential pressure and vacuum at the jet point, allowing oil, water, and gas to comingle with jet fluid and lifting oil, water, and gas up the surface.

The jet pump consists of several major components.

- **The surface equipment consists of the following:**

  - Prime mover (electric motor mounted on a pump skid)

  - Single acting fluid pump capable of delivering a good rate of fluid and 5,000 psi (the prime mover motor is mounted on a skid)
  - Horizontal water reservoir (holds a constant volume of clean fluid)
  - Fluid filtration equipment such as de-sanders and desilt pots
  - Injection steel flow lines
  - Discharge steel flow lines
  - Bypass lines from power fluid to the well

- **The downhole jet pump consists of three simple parts:**

  1) Jet pump housing (with built-in check valve below the tool)

  2) Jet pump carrier
  3) Isolation packer (Arrow set 1X or double grip)

1. Jet pump housing

The jet pump housing

is made of a short solid piece of steel tube machine-cut and bored in a specific shape and size to meet or exceed lifting requirements. The jet pump housing

is run on the tubing joint, spaced, and set above the isolation packer. The pump housing is threaded on both ends to fit various tubing sizes. The bottom section of the jet housing is designed to fit a steel cage loaded with balls and seats, referred to as the "ball and seat" assembly or the "valve" section. The purpose of the balls and seats is to allow the formation fluid to enter the pump intake housing and also prevent the loss of fluid back into the open perforations during fluid injection. The valve assembly is locked with a special ring at the bottom of the threaded sub.

2. Jet pump carrier

The carrier bar plays a major function in fluid displacing. The jet carrier consists of several pieces as a whole:

a) The fishing neck on the top is to latch on, fish, and pull the carrier bar out of the well for repair or replacement.
b) The fluid intake hole is to pump fluid down the tubing string and out through the hole.
c) The actuator tub is to hold the disk in position (there are different sizes of disks).
d) The disk is used to reverse-circulate the fluid out of the carrier bar to the surface.
e) The carrier body is to hold the jet nozzle in position.
f)   The Viton rings are used to seal off the bottom of the jet pump.

3. Isolation packer

The main purpose of the isolation packer is to prevent the annulus fluid from falling back into the open perforation (the packer is used as a fluid saver). Make certain that the production casing string is in good condition before using a hydraulic jet pump. The casing MIT test is necessary in jet pumping application. A good-condition casing string above the packer is necessary to lift the fluid efficiently (a jet pump will not work in a well with casing leaks).

Jet Pump Running Procedure

- Move in and rig up the workover rig.
- Nipple up and test the BOPs high and low as required.
- Inspect and lower the isolation packer in the well (check all the elements).

- Install a 4′ or 8′ IPC pup joint on top of the isolation packer to space out, pick up, and lay down the tool.

- Pick up and make up the jet pump housing on one tubing joint. Then make up the jet pump housing on the 8′ sub onto the isolation packer. Make up the pump housing by hand to avoid tong damage to the pump housing area (avoid wrench damage at or near the orifice). Use a wrench flat to avoid any damage to the housing.

Trip in the hole, testing the tubing string as required. Make up the tubing joints by hand first; then use hydraulic tongs

for the final makeup. Apply a light thread compound on the tubing pin ends only.

- Continue in the hole with packer slowly.
- Finish tripping in the hole while drifting and testing the tubing string going in the hole as required.
- Trip in the hole with production equipment to the landing depth.
- Space out the packer at the setting depth; land and set the packer as required.
- Fill up the wellbore with water (casing and tubing); test isolation packer as required (ensure that the packer is holding pressure to avoid fluid leaks).
- Check the well for flow.
- Nipple down and remove the BOPs.
- Nipple up the wellhead and install all the flow lines (injection line and return line).
- Rig up the surface equipment and fill up the casing and tubing with a sufficient volume of clean fluid. Circulate the well down the tubing and out of the casing string.
- Circulate the well properly using clean well fluid and/or fresh water.
- Once the hole is filled up with the circulated fluid, test the isolation packer (if necessary) and shut down the fluid pump at the surface.
- Remove and fully open up the crown valve and bull plug on the top connection on the tubing string. The tubing string must be full of static fluid.
- Drop off the carrier bar completely and allow the drop bar tool to fall to the bottom at the seat (the tubing should be full of fluid).
- Wait on the tool to reach the seat to avoid jamming the tool.
- Pressure up on the tubing string and seat the carrier bar.
- Put the well in pumping mode and check the pump pressure and pump action.
- Continue watching the pressure and fluid circulation at the surface.
- If the pump action and fluid gain are satisfactory, check the surface lines for possible leaks.
- Check the fluid level in the horizontal separator to make sure not to lose the fluid.
- Turn the well over to production. The circulating pump pressure may reach as high as 3,000 psig.
- Check for and avoid all flow line vibration between the hydraulic pump and the wellhead.
- Install pressure gauges to monitor pumping operations.
- Rig down the service rig and move out (the work is complete).

Jet pumping is a very simple and unique artificial lifting method with no moving components at the surface or subsurface. Unlike other various artificial lifting tools and equipment, the jet pump is simple and does not require too much downhole maintenance. It is easy to retrieve the tool for repair or replacement using reversed circulation. No wire line tools or service rig is necessary. The jet pump can be run in deviated holes or horizontal wellbores.

Artificial Hydraulic Pumping Principle

The hydraulic pumping principle is basically the same as artificial jet pumping. The hydraulic pump consists of three major components. The pump housing is made up of special solid stainless steel to hold the standing valve and the pump nozzle assembly in place. The hydraulic pump housing is made up and screwed at the bottom of the tubing string, just above an isolation packer.

Pulling a Hydraulic Lifting and Jet Pump Out of the Well for Repair

- Move in and rig up the workover rig (pulling unit).
- Hold and document safety meetings.
- Install and test the safety anchors (use a base beam if safety anchors are not available).
- Rig up the surface pump and/or fluid pump truck.
- Fill up the casing and tubing string with produce water.
- Reverse the fluid jet nozzle (pump down the casing and out of the tubing string).
- Circulate the tool into the surface valve; quickly shut the crown valve against the tool while shutting down the circulating pump at the same time.
- Bleed off the trapped pressure and remove the retrieving cap.
- Latch on and recover the tool using a special overshot (send tool for repair or exchange).
- Remove the wellhead assembly and flow line manifold.
- Nipple down the wellhead (Larkin head packing and slips).
- Check the wellhead and replace any component parts as needed.
- Nipple up and test the BOPs as required.
- Rig up the floor and handling tools.
- Release the packer and work free (wait ten minutes for the fluid to be equalized).
- Pooh while strapping and scanning the tubing string out of the hole.
- Check the tubing joints for holes and corrosion pits.

Discard any bad joints.
- Finish out the hole with the tubing string and lay down the following components:

  - – Hydraulic pump housing
  - – Sanding valve
  - – Isolation packer (check and report lost rubber elements)

- Send the tools and equipment to the repair shop for maintenance or replacement.
- Make a note of lost packer elements.
- Prepare to tag the well for filling. Wash and circulate the wellbore clean if necessary.

Running Artificial Hydraulic Pumping Tools

Unload and check all used or new hydraulic lifting tools. Rig up the hydrostatic fluid testing unit to test and drift the tubing string.

Tripping the Hole with Artificial Hydraulic Pump Components:

a) Use an isolation packer (arrow set 1-X isolation packer; check the rubber elements and slips) and hydraulic pump housing (make the tool by hand using a pipe wrench on the flat side only to avoid pinching or damaging the pump housing).

b) Install 8' of a new internal coated pup joint above the pump housing as well as X number of tubing joints (the tubing joints must be fully tested and drifted for the standing valve and jet nozzle to pass through with ease and land in the hydraulic pump housing).

c) Check the well for flow and circulate to kill the well if necessary.

d) Nipple down and remove the BOPs.

e) Space out and set the packer as deeply as required (it is preferred to use and set a three-element tension packer to avoid leaking).

f) Install the pack-off and wellhead equipment.

g) Install the flow line manifold to the hydraulic unit.

h) Inspect, check, and line up all the valves properly.

i) Rig up the fluid pump (rented pump truck).

j) Fill up the wellbore and the hydraulic vessel. Break the circulation down the tubing string and out of the casing string using a pump truck if necessary.

k) Drop the standing valve and allow the tool to reach the bottom.

l) Close the surface casing and test the isolation packer and casing string at 1,000 psi or as required (the casing string and packer must not leak).

m) Bleed off the casing pressure.

n) Remove the crown cap on top of the Christmas tree, drop the hydraulic jet pump, and wait on the pump to reach the bottom and be seated (the tubing must be full of fluid).

o) Install the crown cap and start pumping the fluid down the tubing string, through the jet pump nozzle,

and out of the annulus between the tubing and the casing string. Once the hydraulic jet nozzle is seated into the standing valve, the surface pump pressure will increase to 2,500 and 3,000 psig.

p) Check the wellhead and all the surface equipment for any leaks.

q) Put the well to production and monitor the pump performance (do not increase the pump pressure to increase the fluid; increasing the pressure will not increase any additional fluid and will cause equipment damage).

r) Pictures!

Advantages of hydraulic Jet pumping:

- Simple run, and evaluate lifting operation
- Low artificial cost
- works well in deviated bore holes
- Good artificial system
- Can lift one hundred(100) to three thousand(3000) barrels of fluid per day
- perform best at depths of 1000 feet to 8000 feet

Disadvantages of hydraulic Jet pumping:

➢ Hole or holes in the housing assembly
➢ Tools failure due to trash and formation solids
➢ Packer leak and/or check valve leak

- ➢ Causing fluid cuts on tubing collars
- ➢ May cause fluid cut and corrosion holes in casing string
- ➢ High injection pressure on tubing string
- ➢ Higher surface equipment maintenance and repairs

ELECTRIC SUBMERGIBLE/ SUBMERSIBLE PUMP (ESP)

CHAPTER V

The electric submersible pump (ESP) is an artificial lifting method that operates with a downhole electric motor/motors only. The system demands and consumes lots of electricity to operate. The ESP equipment is made up and run on a tubing string to the determined wellbore operating depth above the open perforations. A closed-system electric motor is run in a well along with a seal section, a fluid intake, and a closed-system pump/pumps to the desired pumping depth to lift the fluid. The electric power is conveyed from the surface to a downhole electric motor/motors using a three-phase flat or round electric cable

that is fastened to the outside of the tubing string. The electric cable is fastened to the outside of the production tubing string by flat steel bands

at several points on each tubing joint from the surface to the desired pumping depth and to the electric motor down the hole. All the components of the electric submergible pump may be suspended in wellbore fluid (oil, water, and formation gas).

How an ESP Operates

Once the electric power reaches the electric motor, it creates an electromagnetic field, causing the electric motor shaft to turn and rotate all the inner connected steel shafts from the downhole motor up through the electric sub pump/pumps,

causing wellbore fluid displacement upward through the tubing string and up to the surface facilities. ESP artificial lifting pumps are capable of displacing large volumes of clean fluid without sand or solids as deep as 10,000'. Large volumes of fluid (an incredible 16,000 barrels of fluid or more) can be achieved depending upon the wellbore equipment design, reservoir deliverability, and fluid characteristics.

ESP Lifting Components

- Tubing strings of various sizes (2 ⅜" to 4 ½")
- Drain sub (useful but optional tool)

- Check sub (useful but optional tool)
- Electric cable (either round or flat)
- Electric submersible fluid pump/pumps of various sizes and stages
- Fluid intake or gas separator
- Seal section
- Downhole electric motor/motors of various sizes and horse power levels

The ESP equipment is designed in a semi-closed system of special steel-cased tubes of various sizes to fit through various casing string sizes.

Pulling and Running Procedures of Submergible/Submersible Pumps

Pulling ESP Procedure

- Check and observe the overhead electrical power lines to the location.
- Spot and use a base beam or drive permanent safety anchors on location. Test the anchors to 30,000 lbs.
- Move in and rig up the workover rig (pulling unit).

- Spot the rig pump and clean the fluid tank. Fill the tank with clean salt water.
- Disconnect the electric power at the junction box and to the well.

- Use the safety LOTO procedure when it is necessary.
- Check all the electrical fuses and wires inside the main control boxes. Make repairs and improve connections if needed.
- Move in and spot the cable spooler and related tech man equipment.

- Check the ESP cable and downhole equipment for problems.
- Check and record the shut-in casing and tubing pressure.
- Bleed off the tubing and casing pressure slowly into the rig tank.
- Check the tubing and pump landing record before you nipple down or nipple up using special ESP BOPs.
- Drop the shear bar to shear off and break the shear pin, if any (need to check). Shearing the shear pin will allow you to circulate and avoid pulling the tubing string and getting the downhole equipment wet.
- Circulate and kill the well if necessary to keep the wellbore fluid static for the workover duration (circulating through the pump or shear pin sub as needed). If there is no shear sub in the well and you are unable to circulate the well bullhead, use heavy fluid if necessary to kill the well (the fluid must be static while tripping the tubing and submersible equipment at all times; do not take the chance of pulling the ESP equipment in a live well).
- Back off the bolts and nuts and remove the wellhead packing elements. Pull out the landing slips carefully to avoid damage to the electric cable and wires.
- Clean up and store all the wellhead parts in a clean bucket of diesel.
- Check the wellhead packing and replace with new equipment if needed.
- Pick up on the tubing string slowly and carefully; clear to remove landing slips (do not hammer down on the wellhead to jump and free the wellhead slips out of the wellhead bowl).
- Avoid losing any slip segments or objects in the hole. Losing slip segments in the annulus will cause expensive fishing.
- After removing all the packing pieces and slips from the wellhead bowl, prepare to check and test the cables again at the wellhead (the problem may be at a pinched point or a damaged cable at the wellhead).
- Test and examine all three wires (three legs

for damages that might cause downhole equipment failure. Any visible cuts or damages to the electric cable wires might cause ESP failure. Recut and repair the wires if any.

- Put the well back on pumping mode without pulling the ESP equipment.
- If retesting the cable indicates downhole problems, proceed.
- Pick up on the tubing string slowly and carefully (watch the tubing weight). Pull and remove the wellhead bowl and flange.
- Install the wellhead flanged adapter to rig up the BOPs.
- Install and nipple up a special BOP stack for the submersible cable

and equipment (use special BOPs to close the wellbore while tripping a round or flat cable). Always be prepared for a possible well flow emergency while tripping the tubing string and cable assembly.

- Latch on the tubing string while calculating the bottom hole assembly weight.
- If the tubing is parted, shut down the operation to evaluate the well problem. Do not pull on the parted tubing with the attached cable (this will cause the cable to snap, loop, and ball out inside the casing tubing annulus).
- If the tubing or submergible equipment becomes stuck, avoid over-pulling on the tubing string. Shut down and evaluate the wellbore condition.
- Never over-pull on the tubing with the attached electric cable. Excess pulling on the tubing string will stretch and break the electric cable and will result in costly fishing work.

Why ESP Equipment Gets Stuck

a) Sand, scale, and drilling mud bridges around the tools
b) Dropping objects from the surface on the cable and pump
c) Parted tubing

d) Parted cable and tubing string
e) Tight casing spots or shifted casing
f) Holes in the casing string releasing mud and solids

- If the tubing string is free, prepare to install the BOPs (check the string weights).
- Drop a weight bar to shear off and drain the fluid inside the tubing string. The tubing string contains oil, gas, water, and suspended solids. Shearing off the pin on the shear sub in the tubing will prevent wet tubing strings and may avoid accidents and environmental pollution.

- Dropping a 1″ × 7′ solid weight bar may shear off the pin in 2 ⅞″ and/or 2 ⅜″ tubing to drain off the fluid before pulling the production string.

Possible Problems from Breaking the Shear Pin

a. The weight bar is too light (not enough weight to knock out the shear pin). You may have to drop a heavier bar to knock out the pin.

b. The fluid may be of heavy density (thick oil or mud slowing down the bar tool).

c. The weight bar may be small in diameter. The small-diameter bar may pass and land on the low side and may miss the pin.

d. The shear pin may be sanded up or have trash above it.

e. The shear pin is knocked out, but the tubing is wet. If the casing and tubing are full of fluid, the liquid will be too slow to fall (pulling the tubing slowly will allow the oil and water to fall).

f. The shear pin is broken, but the tubing is wet. The bar or cable may be blocking the hole at the shear pin hole; the fluid will drain slowly as the tubing is pulled out (pull the pipe slowly if the well is full of fluid).

g. You may not have a shear sub on the tubing (dropping a bar may damage the pump). Do not perforate (may cut or damage the cable).

- If the tubing string is free with a normal string weight, continue out the hole.
- Pick up extra tubing joints. Band the excess cable to the tubing string and lower down the tubing string. This will assist you as you nipple up the BOPs safely and quickly without pushing and pulling the cable through the BOP (also to close the rams easily and safely).
- Install the BOPs and handling tools and equipment to pull the tubing string.
- Rig up the spooler sheave up to the derrick leg at the correct angle and direction to spool the cable out (avoid rotating the tubing and twisting the cable for any reason).

- Avoid oil and water pollution while spooling the cable from the well to the spool unit (you may use plastic sheets and boards to keep the cable from dragging on the ground and cause oil spills).
- Pooh with the tubing string and electric submergible cable while cutting the steel bands off the cable/tubing (avoid dropping any steel band in the hole; keep all the steel bands together in a plastic bucket).
- Before the ESP equipment reaches to surface, shut down and observe the well flow (if the well starts flowing, you cannot shut the BOPs on large ESP equipment; you may pump kill fluid to be on the safe side).
- If there is no flow, pull the tubing to the surface, and lay down the drain sub,

steel bar, and check sub. pump heavy fluid in the annulus if necessary (Safety first!)

- Pull the pump assembly to the rig floor with a fishing neck looking up above the rig floor. Start breaking down the ESP equipment (the ESP lay-down procedure may take some time). Make sure the well is dead. Several sizes of special circular pipe clamps with inserts, an "old-man stand" tool,

and lift chains are furnished to pull or run the ESP equipment safely and efficiently. The special circular clamps are used to lift and lay down ESP equipment to avoid damages to the equipment and prevent tools from getting lost in the well.

- Pull the ESP equipment slowly to the rig floor. Install the special clamp and lifting chain assembly. Set and rest the pump assembly on the old-man stand.

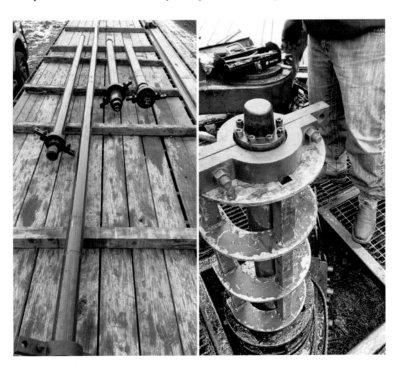

- Back off and lay down the top landing joint. Get ready to pull and lay down the ESP equipment in singles.
- Rig up the technician equipment, lifting tools, and combo clamp. Pick up on the pump above the rig floor using rig blocks. Cut and remove steel bands from the pump/pumps and motor.

- Test the pump shaft rotation and oil. Remove and lay down the pumps using ropes and cat line (a good pump shaft should rotate easily).
- Pull and test the fluid intake (gas separator). Remove and lay down the fluid intake.
- Pull and test the seal section. Check the shaft rotation and check for the quality of oil inside the seal section (the seal should be free of water)
- Continue to pull and test the electric motor. Check the oil, rotation, and lead cable and test all the components for malfunction.
- Lower the spool sheave to the rig floor and hold onto the cable before cutting the cable to avoid the cable from jumping out of the sheave and causing injury.
- Cut the cable at the motor lead, pull the cable by hand, and rig down the spooler wheel

and safely place it on the ground.
- Pull and lay down the electric motor.
- Strip and check all three legs on the electric motor.
- Keep records of field testing information on all the equipment that was pulled out of the well, including models and serial numbers of all the equipment.
- Load up all the electric submergible equipment. Send the equipment for testing and repairs if necessary.
- Trip in the hole with ten stands of tubing and close the wellbore in.
- Wait on the new ESP or tested equipment to return.

Causes of Downhole Submergible Equipment Failure

a) Burned electric motor

b) Water in electric motor
c) Locked-up motor caused by sand and scale getting into the system
d) Cable breaks or burned wires
e) Holes in tubing string
f) Parted tubing string
g) Seal section failure (with water and no oil)
h) Pump failure (sanded-up pump)
i) Loose parts inside the pump
j) Objects lodged inside of pump/pumps
k) Burned surface of electric wires
l) Single phasing electric power

ESP Running Procedure

– Several sizes of special circular pipe clamps and an old-Man stand are furnished to run the ESP equipment safely and efficiently.
– The clamps are used to lift and lay down ESP equipment to avoid damages and keep the equipment from getting lost in the well.

– All the components of the electric submergible pump are made up and flanged up together using six high-strength screws on each piece of pump equipment.

How to Run ESP Equipment

A high-strength chain attached to a special-made clamp is used to pick up, hold, and run the ESP equipment.

1. Safety first (make sure the well is dead) Install the correct-sized circular clamp on the electric motor. Pick up the electric motor with rig blocks.
2. Lower the motor in the well and set on the old-man stand. The protective end caps on top of the motor will be removed. Old oil that was put in at the repair shop will be drained out, and the shaft rotation will be checked. The void space will be filled with fresh and clean special oil.
3. Pick up the seal section with the pipe clamp, remove the end caps, check the shaft rotation, drain any old oil, and replace all the old O-rings.
4. Flange up the seal section on the electric motor and tighten the flange bolts. Do not over-torque the flange bolts. Lower the seal section and motors into the well. Set and rest the seal section on the old-man stand.
5. Pick up the gas separator (fluid intake) and flange up on the seal section. Check the ports, shaft, and seals.
6. Pick up and install the fluid intake (gas anchor assembly.

7. Pick up the fluid pump/pumps and remove the end caps. Check the shaft rotation on each pump, change any old oil, replace old O-rings with new O-rings, and flange up. Lower the pump/pumps in the well.

8. Pick up and install the last pump as directed above. Check all shaft rotations using a pair of pliers; recheck all the flanges (do not over-torque the bolts). Once the top pump (the last pump) is flanged up, check the "bolt-on head" and the crossover to make sure the makeup threads are not flat or galled because of continuous work.

9. Pick up the entire ESP equipment assembly (motor, seals, gas separator, and pumps) above rig floor level.

10. Start from the bottom of the ESP equipment to check, drain old oil, and fill up the equipment with new oil.

11. Recheck and drain out any old oil and fill up the electric motor with new fresh special oil using a hand pump. Continue injecting fresh lube oil until all the old oil and air bubbles run out from the top ports on the electric motor. Close the ports and seal them tight after lubrication.

12. Lower the motor in the hole; lower the seal section to the rig floor. Check and fill up the seal section with new oil until clean and clear oil pours out from the upper port. Close the ports on the seal section after lubrication.

13. Wipe off and clean up the oil and slack off in the well with the motor and seal section. The electric motor and seal section must be full of special cooling/lubricating oil at all times to avoid overheating and causing electric motor problems.

14. Lower down and check the intake and pumps. Check the pumps and shafts to make sure the shaft is turning free on the intake and fluid pumps.

15. After all the ESP equipment is lubricated and checked, pick up the ESP assembly on the rig floor again, whip off the electric motor dry, and install the motor lead extension (MLE) from the pump and up the hole. The MLE is an important connection where

the electric power reaches the motor down the hole. The cable wire endpoints must be cleaned with a knife and/or sand paper to remove the varnish coats for contacts.

The connection points will be wrapped with special electric tapes and epoxy to keep water from entering the motor through the MLE. After the MLE is installed properly, apply a special epoxy to cover the entire area and wait at least thirty minutes for the epoxy to harden.

16. After the MLE is installed, hold the lead cable flat against the motor body and start banding with a flat steel band.

Several steel bands may be used to secure the cable straight and tightly (band each 3' spacing on submersible equipment).

17. Lower the ESP equipment and apply steel bands on the seal section, the gas intake, and all the pumps going down the hole. Use several bands on the ESP equipment at least 3' apart to secure the cable as flat against the pump as possible (do not rotate).

18. Install a 6' pup joint on top of the pump (the bolt-on head or fishing neck) to latch the blocks on and to remove the chains and clamps using the rig elevators.

19. Trip in the hole with the following ESP assembly equipment from the bottom up:

- Electric motor or motors
- Seal section
- Fluid intake (gas separator)
- Pump/pumps
- 6' pup joint to lift ESP equipment
- Two full tubing joints (one stand)
- Check valve
- Two full tubing joints
- Drain sub
- The rest of the tubing string up to the surface and to the wellhead

The application of a shear drain sub and fluid check sub is recommended in the installation for various practical operation purposes. The drain sub is used to shear off a pin inside a tubing sub by dropping a weight bar to drain the fluid out of the tubing string before pulling the tubing. This will prevent wet tubing strings as well as oil, gas, and salt water pollution and danger to rig workers. Shearing off the shear pin will help circulate the well if necessary. The check valve is used to keep the fluid from falling back in case of shutdown. This may prevent reverse rotation and shaft twisting in the pump because of the hydrostatic column of fluid in deeper wells.

20. Apply a steel band every 3' on the motor lead up to the top of the sub pumps.

21. Continue going in the well with the ESP equipment and tubing. Use one steel band every 3' on the first four joints above the pump.
22. Install one steel band 3' below each tubing collar and another steel band 3' above each tubing collar. Using one steel band in the middle of the tubing joint all the way up is an

option. While running and banding the cable in the well, the electric cable must be pushed and held straight and tightly against the tubing body before the banding cable must be held in the same direction without turning the tubing all the way to the surface (avoid twisting and wrapping the cable around the tubing).

23. You may test the tubing string while going in the hole to drift and detect collar leaks

or finding holes in tubing joints.

24. Avoid rotating the tubing going in the hole or coming out of the hole to avoid twisting the cable around the tubing string and damaging the cable.

25. Avoid corkscrewing and galling the pipe and collar threads. This will cause you to lose ESP equipment in the well and create costly fishing work.

26. To avoid galling threads, apply light lubrication on the pin ends and make up the tubing by hand for at least three rounds to find the starting threads before applying tubing tongs to apply a required makeup torque.

27. Use a thread dope/compound on the pin ends of the tubing only. The BESTOLIFE thread compound is preferred.

28. While going in the hole, pay attention so as not to crush the electric cable with the slips. The cable must be cleared to one side before closing the pipe slips

every time.

29. Remember that the starting thread is on top of the collar and not four threads down into the collar.

30. Once you are finished going in the hole, stop and check the well for flow. Kill the well if necessary. To remove the BOP stack safely and quickly, go to the next step.

31. Pick up an extra tubing joint and strap the excess electric cable. Lower the tubing with the cable in the hole and start to nipple down the BOPs. This will help remove the BOP stack without pushing and pulling on the electric cable in and out of the BOPs.

32. Install the wellhead and flange up. Pull and lay down an extra tubing joint and excess cable to the surface.

33. If spacing and splicing is needed, make sure not to cut the cable short.
    • Make up a special socket flange head on the landing joint.
    • Install the cable head tightly by hand.

    • Pick up on the tubing string and short cable until you reach the long cable coming from the well.
    • Once the two cables meet, add 3" and cut the long cable.
    • Remove the protective coating, clean up the wires with a knife and sand paper, and wrap with five layers of special electric tape as best as possible.

- Wrap with armor as flat as possible. Do not use a hammer to make the cable wrap flat.
- Install the socket in place and secure the cable with steel bands. Use heavy bands on the landing joint (one band every 5′).
- Lower the landing joint and nipple up the wellhead flange.

34. Check and connect the cable to the electric panel and boxes.
35. Extend and install the cable wires from the wellhead to the junction box.
36. Install the flow lines. Make sure all the flow line connections are made up tightly to avoid shutting the well in because of flow line leaks.
37. Check all the flow lines to make sure the valves are fully open.
38. Secure the electric cable to the wellhead and put the well on pumping mode.
39. Check fluid flow and pump rotations if necessary.
40. Rig down and move out the service rig.
41. Clean up the location.

Head Curve (multifrequency)

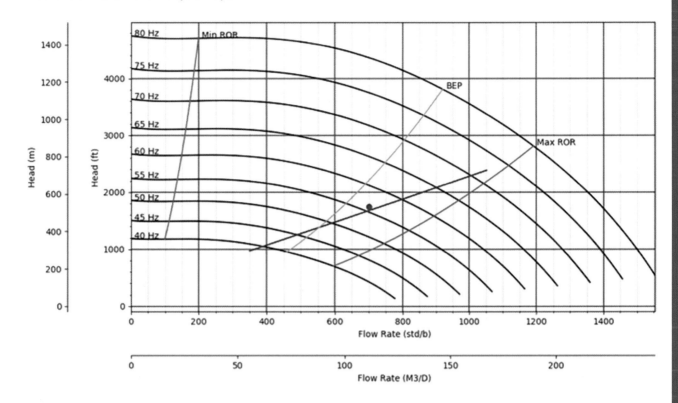

Head Curve (multifrequency)

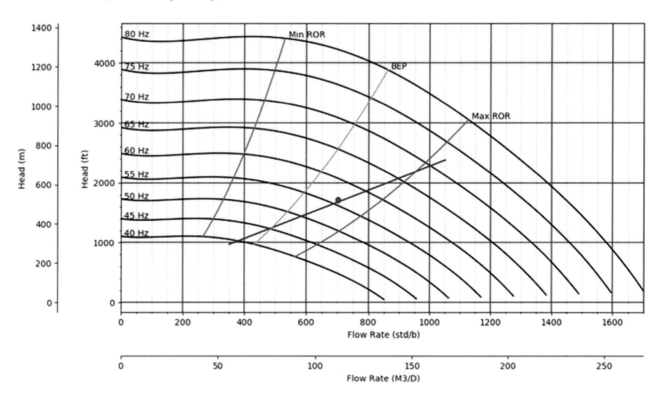

Individual Pump Curve

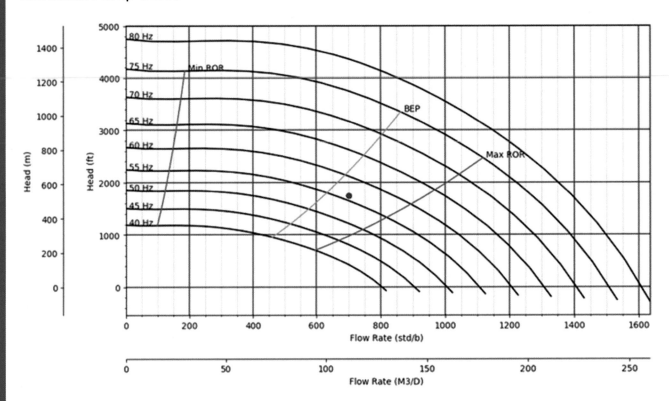

Individual Pump Curve

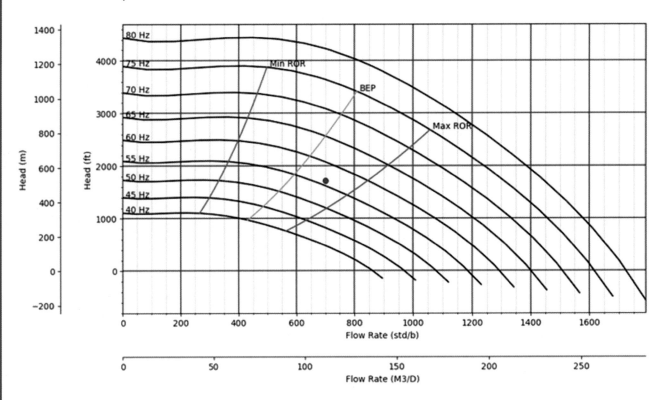

BHP Curve (multifrequency)

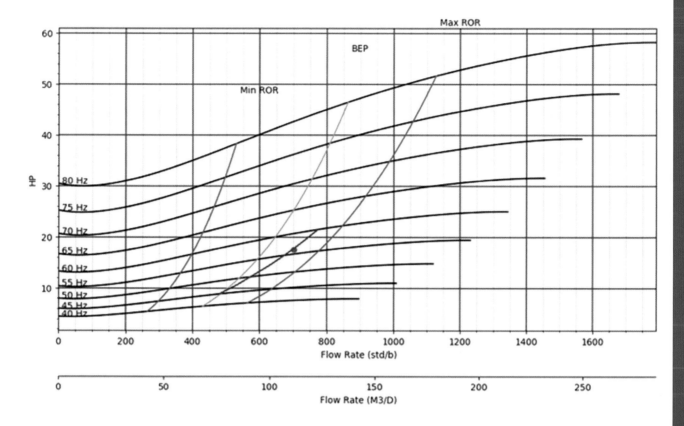

BHP Curve (multifrequency)

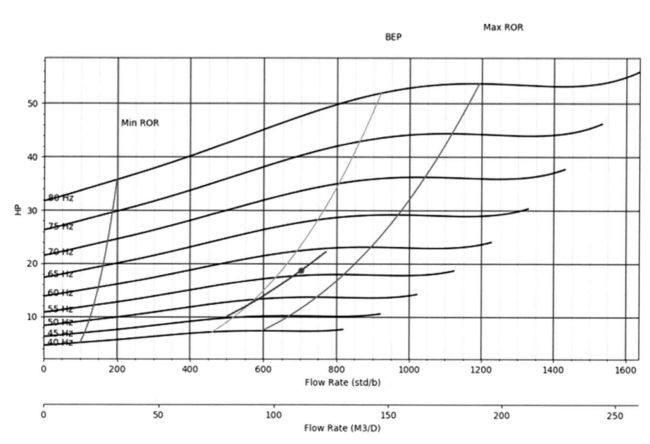

Pump Curve

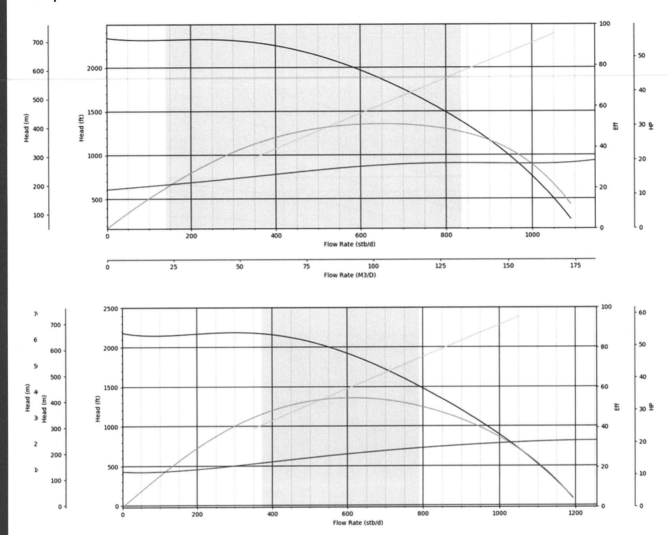

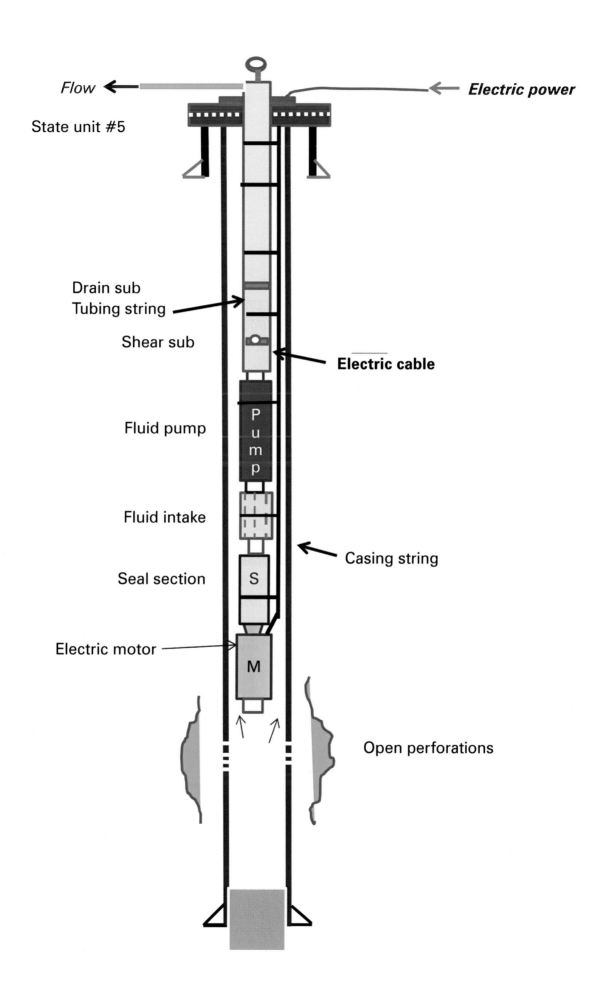

Flow ← ⟵ *Electric power*

State unit #5

Drain sub
Tubing string

Shear sub

Electric cable

Fluid pump

Fluid intake

Casing string

Seal section

Electric motor

Open perforations

Components of electric submergible artificial lift:

    a) Electric motor/motors
    b) Seal section
    c) Fluid intake
    d) Electric motor/motors
    e) Electric cable (flat or round cable)
    f) Check valve (option)
    g) Shear sub (option)

Advantages of Electric Submersible artificial lift

- *Environment friendly artificial lifting system*
- *No moving parts surface and subsurface compare to beam pumping*
- *successful Lifting range of 2000 feet to 10000 feet*
- *Known as best high artificial lifting method ranging 200 barrel s to 20,000 barrels of fluid per day*
- *Easy to supervise and manage the lifting operations*

Disadvantages of Electric submersible artificial lifting

➢ *Consumes too much electricity*
➢ *High cost of initial installation*
➢ *Higher cost of equipment repairs compare to other lifting system*
➢ *Require clean fluid and low gas/liquid ratios*
➢ *Will fail in presence of sand, scales and solids*
➢ *High fishing cost of 200/foot when cable and tubing become parted*

➢ *Difficult to fish parted tubing string and cable*
➢ *Will cone water drive in reservoir*

VI
CHAPTER

PRINCIPAL OPERATION OF PROGRESSING CAVITY PUMP

The PCP is basically a rotary-type positive-displacement pump. The PCP is a simple designed artificial fluid pump that can be run and lift fluid in any wellbore at shallow depths ranging from 1,000' to 4,500' based on pump size and the following pump stages:

$$\begin{Bmatrix} 5 \text{ stage } 2 \\ 9 \text{ stage } 2 \end{Bmatrix}$$

Generally, 5 stage 2 and 9 stage 2 pumps are generally used for low production, ranging from 20 to 45 barrels per day in shallow wells ranging from 1,000' to 2,000'.

$$\begin{Bmatrix} 5 \text{ stage } 3 \\ 9 \text{ stage } 3 \\ 18 \text{ stage } 3 \end{Bmatrix}$$

In addition, 5 stage 3, 9 stage 3, and 18 stage 3 pumps are used in wells with depths of 1,000' to 3,000' lifting 100 to 150 barrels per day. The 9 stage size 5 pumps and 18 stage size 5 pumps are heavy-duty pumps. They may be applied in wells with depths of 2,000' to 4,500' lifting 700 or more barrels per day.

The progressing fluid pump is a simple pump that consists of only two major subsurface components:

a) The Stator

The stator appears as a smooth heavy-duty pipe with a built-in double-helix element internally. It is precision-molded with a tough double-corrosion-resistant synthetic elastomer compound that is permanently glued and bonded inside the steel pipe housing. The stator assembly is run and suspended on the tubing string, similar to the tubing pump on the artificial beam pumping. The stator size may range from **2 ⅜″ to 3 ½″** with various lengths.

b) The Rotor

Rotors are made and manufactured with high-strength, precision-machined solid bars. They are made of solid chrome-plated steel bars with single helixes (**curvatures**). The rotor assembly is run on a steel sucker rod string similar to the pump plunger of tubing pumps on the artificial beam pumping. Rotor rods may come in different sizes and lengths. The steel sucker rods used in the progressing cavity and rotary motions are subject to constant torque forces. The selected rod string must meet or exceed the applied torque requirements during the operation.

There are several torque components applied to the rod string:

I) Hydraulic head torque (total dynamic head)

    1) Lifting head (pump depth)

2) Wellhead pressure head
3) Line friction

II) Rotor and stator rotational friction torque
III) Torque caused by fluid viscosity and solids abrasion

Progressing Pump Range of Application

The progressing pumps are effective and efficient for depths of 1,000' to 4,000', lifting 20 to 500 barrels per day based on the pump size, fluid density, and reservoir drawdown factor.

Speed Control Factor

The pump speed can be controlled by changing the sheave or VSDs, similar to artificial beam pumping or submersible lifting.

Operation Troubleshooting

Like any other artificial lifting system in oil field operations, the progressing pumps are subject to surface and subsurface mechanical and lifting problems. Generally, if the drive head turns the rod string assembly with ease but without lifting fluid, it could indicate one of several downhole problems, such as parted rods,

parted tubing, and worn rotors

and/or stators.

The excessive increase of bottom hole temperature and some chemical reactions on the stator and rotor surface can cause significant mechanical change on the tools. If the drive head becomes significantly difficult to rotate with high torque, it is an indication of excessive abrasion or jamming the rotor inside the stator. Unclean wellbore fluid such as sand and mud may create unwanted fishing work down the hole. An inaccurate tubing tally and bad rod spacing may cause the well not to lift fluid properly as well.

There are basically two major diagnostic problems associated with progressing pump artificial lifting:

I) The pump is running but not lifting fluid.

- Touch the unit with the back of your hand (safety measure).
- Reach out and turn off the unit. As soon as the unit is turned off, the drive shaft should stop turning immediately (often with a slight back spin). If the drive shaft does not stop immediately and spins for several turns before coming to a complete stop, this is an indication of several downhole problems.
- Manually turn the drive shaft using a pipe wrench to check the RPM and torque. If the drive shaft manually turns, then this could be one of the following problems:

    a) Parted rods

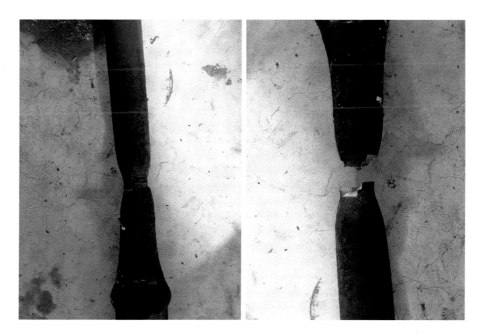

    b) Parted tubing string
    c) Backed-off stator

d) Broken rotor

II) The well is not lifting fluid, and the unit is not running.

Check the electric power to the unit and stay clear from the stored torque (the torque may cause back-spinning). The problem can be any of the following scenarios:

- Bad spacing or rod stretch (the locked drive shaft could be due to the rotor resting on the stator stop pin (the crossover swage)
- Sanded-up rotor and stator (pump stuck)
- Sanded-up tubing string (stuck rods and pump)
- Object around the rods (broken rod guides)
- Frozen motor
- Frozen drive head bearings

Below is a step-by-step procedure to diagnose pump problems:

I) The unit is running, but no fluid is flowing.

a) Turn the unit off.
b) Manually rotate the drive shaft.
c) If the drive shaft is turning, remove the drive head.
d) Kill the well; make sure the well is dead.

e) Pull the parted sucker rods. If the rods are parted, fish the rods out.

f) Replace the bad rods and check the pump rotor.
g) Run the rods and rotor in the hole.
h) Space out.
i) Install the drive head back on.
j) Turn the unit on pumping mode again.
k) If the unit is lifting, put the well on test mode.
l) If the well is pumping, rig down and move off.

II) The unit is running, but no fluid is flowing.

a) Turn the unit off.
b) Manually rotate the drive shaft.
c) If the drive shaft is freely turning, remove the drive head.
d) Kill the well; make sure the well is dead.
e) Pull the sucker rods and rotor.
f) If the sucker rods and rotor are okay, kill the well with kill fluid.

g) Prepare to pull the tubing string with stator!

h) Nipple up and test the BOPs.
i) Latch on to pull the tubing and stator.
j) If the tubing is parted, fish out the tubing and stator.
k) Clean up the wellbore.
l) RIH (run in hole) with a new stator on a good tubing string in the hole.
m) RIH with rods and a rotor.
n) Space out carefully (twice).
o) Install the drive head.
p) Turn the unit on.
q) If the unit is lifting, put the well on test mode.

III) The unit is running, but no fluid is flowing.

a) Turn the unit off.
b) Manually rotate the drive shaft.
c) If the drive shaft is not turning (stored torque energy), carefully remove the drive head.
d) Readjust the rotor and stator alignment and spacing.
e) Install the drive head back on again.
f) Turn the unit on to see if it is turning and lifting fluid.
g) If the unit is not lifting fluid, turn off the power to the unit.
h) Remove the drive head again.
i) Kill the well.

j) Pull the rods and inspect the rods and rotor.
k) Kill the well and pooh with the tubing string, looking for holes in the tubing string and a bad stator.
l) Check and replace the stator.
m) RIH with the tested tubing and new stator.
n) RIH with the rods and new rotor; space out properly
o) Install the drive head and put the well on pumping mode.
p) If the unit is lifting fluid, put the well on test mode.

The PCP is basically a rotary-type pump of positive fluid displacement pump. PCPs may be called different names in oil and gas operations:

- Moyno pumps
- Rotary pumps
- Cavity pumps
- Econo lift pumps

The PCP is a simple positive-displacement pump designed to pump a variety of fluids such as chemicals, salt water,

and viscous oils, with low concentrations of formation sand or scales.

The unit will offer lower operating energy requirements compared with conventional beam pumps or electric submergible pumps. The change of production rates can be adjusted by changing the sheave or adjusting variable speed-controlled devices at the surface (an option).

Progressing Cavity Pumps

Like the conventional pumping unit,

the PCP also consists of surface equipment and downhole or subsurface pumping components.

The surface equipment of the PCP consists of the following:

- Drive head
- Prime mover (electric motor)

The drive head is the most important part of the pumping system and consists of the following:

- Cast housing frame
- Drive shaft
- Drive head–body connection
- Thrust bearing assembly,
- Seal and packing elements
- Mounted prime mover (electric motor)

The drive head must meet or exceed the total trust load generated from the bottom of the rotor to the drive head. The trust load depends on the following:

- Revolutions per minute (RPM)
- Generated torque
- Pump size
- Tubing size
- Rod string size
- Fluid properties
- Total dynamic head

The trust load and total dynamic head are the basis in the selection of a particular-sized pump for a particular wellbore. The expected surface and subsurface thrust load forces on a drive head consist of the following:

- Torque and friction caused by high RPM from the surface to the end of the rotor
- Constant torque caused by the friction and abrasion of viscous fluid and solids from the end of the rotor to the surface and up the flow line
- Weight and rod string friction (consists of the rotation of heavy rods under load with sand, solids, and liquid)
- Surface friction (choke and back pressure friction of Ls and Ts as well as back pressure of production components at the vessels)
- Wellhead and back pressure of dump valves (flow line chokes).

The drive heads are either "vertical drive heads" of different models for lower and/or higher fluid rates or right-angle gear-type drive heads for deeper wells with higher RPM. The PCPs are designed to operate efficiently from a shallow range of 500' to 4,500'. The application of PCPs in deeper wells may cause surface and downhole mechanical failures of various types. PCPs are efficient at shallow depths with clean fluid, capable of lifting 50 to 5,000 barrels per day based on the downhole equipment and reservoir deliverability.

The subsurface components of the PCP consist of two major parts:

- Stator tube/housing
- Rotor shaft or rotor bar (helical chrome shaft)

Stator

The stator is run at the bottom of the first joint of tubing going in the well. (Using other tools such as a bull plug, a mud anchor joint, a perforated sub,

and a TAC below the stator is a practical option.) The stator tube is made of API tubing with an internally molded double internal hard helix synthetic elastomer material.

The synthetic material of the stator is permanently glued or molded inside of the heavy-walled tubing housing with accuracy and precision. The internal synthetic material of the stator is durable and can resist friction, abrasion, hostile oil gas interference, and the heat created by the wellbore fluid (the stator internal element is found to be tough). The stators are cut and designed using 2 ⅜″ OD to 3 ½″ OD tubing, 28″ to 181″ in length.

Rotor Shaft

The rotor is run at the bottom of the sucker rod string (similar to a pump plunger

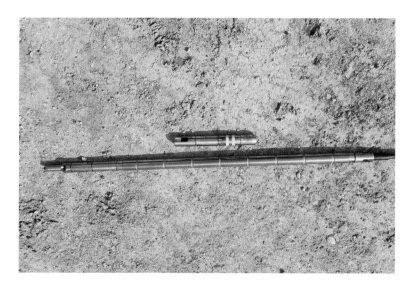

in the conventional tubing pump method). It is made of a solid bar shaped with smooth chrome curves of single helixes. The rotor is a high-strength precision bar that is hounded, machined, and chrome-plated to fit the double helixes in the stator (the rotor must fit the stator for proper engagements).

The rotors are built to be solid and designed in various lengths and sizes, hounded, polished, and chrome-plated as smoothly as possible. The rotors are designed based on the pump size, ranging from ⅝″ to ⅞″ in diameter and with lengths of 29″ to 186″. Each rotor is designed to fit a particular stator (the rotor and stator must perfectly fit together as designed pairs).

How a Progressing Cavity Pump Works

When the rotor and stator

are properly spaced, seated, and engaged in place, a tight seal will develop between the single helix of the rotor and the double helix of the stator. When the rotor turns clockwise inside the stator, the cavities will progress in an upward motion, causing smooth upward fluid displacement. This upward fluid movement will continue until the fluid reaches the surface through the tubing string. The pump output depends on the casing size, tubing size, pump size, pump depth, fluid properties, and surface RPM. The pump output may range from 50 to 5,000 barrels of fluid per day.

Running, Pulling, and Spacing Procedure on PCPs

- Check and remove all the hazardous and unsafe objects from the well location (fill up the rat hole and muse holes on the surface).
- Install the electric panel for the pumping assembly. You may use a portable electric generator if necessary.
- Install and test the four safety anchors on the location. Test the safety anchors at a tension of 30,000 lbs.
- Check the overhead power lines before moving a workover rig.
- Use a base beam as safety anchor equipment if necessary (homemade auger anchors are not recommended).
- Move in and rig up the service rig (workover rig).
- Level off the rig and hang blocks at the center of the wellbore (use a base beam if necessary).
- Open the well; read and record the tubing and casing pressure (bleed off the well pressure in a rig tank).
- Kill the well if necessary (know the well's past operation and workover history).
- Pooh with production equipment, if any.

Wellbore Cleaning and Preparation before Running a PCP

- Tally the tubing string and trip in the hole with a rock bit and casing scraper (keep an accurate tubing tally).

- Use a work string if possible for wellbore washing and drilling (do not use an IPC production string for drilling purposes).
- Wash, clear, and drift the casing string to the plug back depth below the open perforations.
- Wash and circulate the wellbore clean with field-produced water (reverse-circulate down the casing and out of the tubing string if possible to avoid pumping solids into the cased hole/perforation).
- Use PH-6 tubing as a work string in deviated or horizontal wellbores if necessary (avoid running full-size casing scrapers in horizontal wellbores).
- Pooh with a tubing string; lay down the rock bit and casing scraper. Keep the hole full properly.

Prepare to Run the Pump Stator and Production Tubing

- Use IPC production tubing to prolong the useful life of the tubing string.
- Use IPC for producing the string. Plastic-coating the OD of the tubing is not a practical application and is not recommended because of tong and slip damages. If the outside plastic coating is necessary, the tubing damages

must be repaired and recoated on location while running the tubing string in the well.
- Tally the tubing string as accurately as possible while going in the hole.
- Measure the tubing string and anything that goes in the well, including the lengths of the TAC, seating nipple, stator, and crossovers.
- Check the stator for dents and bad threads. Do not run a used or damaged stator in the well (it is difficult to see clearly inside a PCP stator).

- Inspect the internal part of the stator for cuts and damages using a flashlight (looking for internal cuts and swelling helix elements).
- Rig up the tubing testers. Hydrostatically test the tubing string above the stator assembly only (it is not necessary to hydro-test the stator assembly). The stator assembly should be tested in the pump supply shop, and no hydrostatic test is needed in the field.
- Space out and use mud joints, a perforated nipple, and a TAC below the stator if necessary (using a mud anchor and a gas anchor is a good practice to catch solids).

- Make up the tubing joints by hand first; then use hydraulic tongs to make up the tubing string as recommended.
- Apply a light thread compound on the pin ends only (make up the tubing tightly but do not over-torque the tubing joints)
- <u>Remember</u>: The rotor will turn and create clockwise torque, but the stator rotation will create counterclockwise torque, which tends to back off the tubing string joints. You may also use a tubing anchor that is set to the right and released to the left. This kind of tool is used on PCP application (using a TAC is an option).

OIL AND GAS ARTIFICIAL FLUID LIFTING TECHNIQUES

323

- If you are running a PCP in a horizontal and/or highly deviated hole, avoid running a TAC to avoid losing springs (it may be difficult to release the TAC with broken centralizer blades in the lateral wellbore).

Tripping and Tubing Landing Procedure

PCP Running Procedure from the Bottom Up the Hole

1. A blind bull plug
2. Two tubing joints as mud anchors "without holes"
3. 4' or 8' perforated nipple or machine-cut slotted nipple (screen is an option)
4. Special mechanical seating nipple with OD and ID threads to screw the gas anchor (dip tube) below the nipple (Use a 20' dip tube below the pump assembly to slow down gas interference. Do not over-torque the tube to stretch the threads.)
5. TAC (Using a TAC with a double spring is preferred. Do not use a TAC in any horizontal wells with high H2S gas levels. On PCP applications, the TAC must be a **right-hand-set** and **left-hand-releasing** tool to stay in the set position during high RPM.)
6. One joint tubing (as a spacer)
7. The tag bar (The old tag bar was a rod or bar, but the new tag bar is actually a crossover sub with perforated holes. The tag bar is run at the bottom of the PCP.)
8. The stator assembly (This is a long heavy piece of steel tube with an internally bonded double-helix tough material. External Monel coating is an option to slow down external corrosion and pits.)
9. 6' internally coated pup joint as a lift sub (to safely lift up or lay down the stator)

Finish testing and tripping in the hole with IPC tubing, using a stabbing guide and a Teflon rabbit, to the required pumping depth (keep accurate pipe measurements for the correct landing depth).

- Check the well for flow and circulate to kill if necessary.
- Remove the rig floor and all the handling tools (clear the area to remove the BOP).
- Check the wellbore for flow before removing the BOP stack.
- Nipple down and remove the BOP stack.
- Install and flange up the wellhead assembly.
- Space out and set the TAC as required. Check the correct spacing for flow lines. Avoid pulling too much tension on the TAC.
- Install the landing slips and related pack-off components and nipple up (check and replace the wellhead packing to prevent oil and gas leaks).
- Install a special fluid pumping tee on the tubing string. Special, high-quality, and unique flanged pumping tees with boxed ends and flanged tops are used as both pumping tees and high-pressure pack-off BOP assemblies on most PCP systems. (The top of the flange is threaded internally for various connections and lifting applications that must be kept in good operating condition.)

Prepare to Run the Sucker Rods

- Rig up the floor and handling tools and inspect the rotor and rod string.

Prepare to Run the Rotor and Sucker Rods

- On new-model PCPs, the drive head and motor are designed with a flanged BOP assembly (easy to remove and easy to install).
- On old-model PCPs, a pumping tee and a hammer union between the pumping tee and the drive head are used.
- Prepare to Run the Rotor with the Rod String
- Set the helix rotor on the flat ground and inspect the threads and chrome body on the rotor from top to bottom.
- Check the chrome on the rotor for cracks, cuts, or dents. Discard rotors with chrome damages (never reuse any rotor bar).
- Count and recount the sucker rods, including the length of the rotor and/or pony rods (the rod is 25' long; the pony rods are 2', 4', 6', 8', 10' long).

- Attach and make up the rotor bar at the bottom of a sucker rod and start in the hole with the rotor and sucker rods (use high-strength snap on rod guides

if necessary).

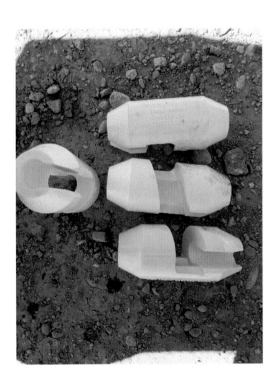

- Make up the rods by hand first; then use hydraulic tongs to make up the rods.
- Use displacement cards to make up the rods as recommended by the sucker rod manufacturing company (use the recommended lubrication on connections using a paintbrush).
- Check all the rods and rod boxes for corrosion pits, cuts, nicks, dents, and wear. Discard any bad rod.

- Trip in the hole with rods to the top of the stator (do not run into the stator).
- Pick up an extra rod and slack off into the stator while holding the rods by hand (you will see and feel when the rotor finds its way, screwing into the stator grooves).
- Rotate the rods using a hand wrench, slowly screwing into the stator. Slack off slowly until the rotor touches the "stop pin" or "tag bar" and stops.

Spacing the Rods

- Make a mark on the rods at the surface. Rotate the rod string with a rod wrench on the stop pin or tag bar for four rounds clockwise until you feel a good torque.
- Pick up off the stop pin while holding the trapped torque. Pull 9" off the bottom and mark the rods again. Continue picking up until the rod unwinds itself (you will feel it).
- Space out with pony rods and a polished rod as needed. Repeat tagging and check the spacing three times to make sure the spacing is correct (do not bump hard on the stop pin).
- Space out off the tag bar as much as necessary to avoid touching the tag bar). If the spacing is satisfactory, pick up and flange up to the drive head.
- Pony rods are purchased in lengths of 2', 4', 6', 8', and 10'. Use the shorter pony rods at the bottom of the top sucker to avoid pulling the rotor in and out of the stator while spacing.
- Install the check belts and sheaves (avoid too much tension on the drive belt to avoid an overhung load). Minimum belt tension without slippage is needed.

- Install and check the electric motor and check for the correct clockwise rotations (avoid counterclockwise motions).
- Check the packing elements and lubrication in the drive head.
- Pick up and mount the drive head and flange up.

- Once the drive head and surface equipment are installed, check and rotate the drive shaft for twenty rounds clockwise to make sure the shaft is turning freely.
- Start the motor and kick the PCP on.
- Check the RPM, and listen to the motor for any abnormal conditions. A start of 200 RPM on the motor is preferred, allowing the stator and rotor to adjust.
- Ensure that all the surface valves are fully open to the tank battery.
- Put the well on test mode while waiting on the fluid to reach the surface.
- Check the surface lines for leaks.
- See the PC unit and equipment pictures.

PCP Pulling Procedure and Required Downhole Equipment

- New rotor
- New stator
- New 16' polished rod with a polished rod box
- New double-bladed TAC to fit the casing size
- New blind bull plug
- Two new tubing collars
- New 20' gas anchor (dip tube)
- New perforated nipple or machine-cut slotted nipple (4' long)
- Few new rods as required
- Few special snap-on or molded rod guides

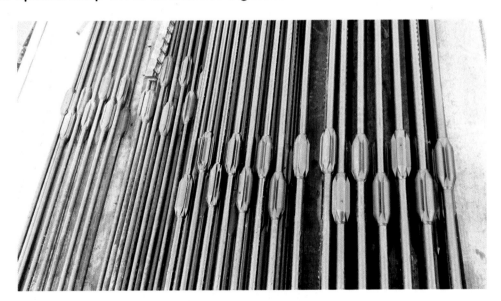

- Pony sucker rods (2', 4', 6', 8', and 10')
- New rod boxes

Pulling Procedure of Rotary Pumps (PCP)

- Move in and rig up the workover rig.
- Bleed off and kill the well properly.

- Screw a short 4' pony rod on top of the polished rod.
- Close the BOP (the fluid pumping tee); tightly screw in the packing elements around the polished rod.
- Latch onto the rods.
- Back off the bolts on the polished rod clamp

and slack off the rods on the pumping tee (the pack-off). The polished rod must not move downhole.

- Make sure to avoid dropping the rods below the wellhead (the special fluid pumping tee must be tight enough and in good condition to keep the rods from going down the hole).
- Remove the pony rod and rod box from the polished rod.
- Install a chain on the drive head, and un-flange the drive head from the pumping tee.
- Get ready to pull the entire drive head.
- Pick up on the drive head; remove it from the polished rod. Spot the drive head with all the electrical cables (place the drive head away from the wellhead and on a stand). You will see the polished rod sticking out above the fluid pumping tee.
- Install the nipple and stripper head on the flanged fluid tee and close shut against the polished rod.
- Back off the rams, open the fluid pumping tee, and check for flow.
- Latch on the polished rod and open the rod BOP
- of the pumping tee. Install the rod stripper to the BOP (pumping tee) and get ready to pull the rods.
- Pick up on the rods and pull the rotor out of the stator. When you are pulling up on the rods, the rotor will cause left-hand rotation while coming out of the stator helix.
- Rig up the floor and handling tools.
- Kill well if necessary using brine water. The well fluid must be static while tripping the rods.

- Pour 10 gallons of diesel fuel down the tubing string to dissolve or soften paraffin around the rods and inside the tubing.
- Start out the hole with sucker rods and the rotor, pulling straight out of the stator assembly. Inspect the rods as they come out of the hole.
- Cut paraffin, if any, and inspect the sucker rods for corrosion pits and damages while they come out of the hole.
- Finish pulling the rods out of the hole. Count the rods, lay them down, and inspect the rotor for damages. Rotor damage

is very obvious (the chrome is gone, cuts on the helix part of the bar, scale buildup, etc.).

Pulling Tubing String and Stator Assembly

- Remove the rig floor and the rod stripper.
- Remove the fluid pumping tee and rod BOPs.

- Remove the wellhead flange etc.
- Slack off and release the TAC, if any.
- Note: The TAC on the PCP system is set and released differently from the beam pumping system. The PCP tubing anchor is designed for right-hand setting and left-hand release. (Slack off and rotate the tubing string by hand for eight rounds to the left; slack off to free. The right-hand-set TAC selection is due to the clockwise rotation on PCP rods to avoid releasing or backing off a left-hand-set TAC anchor.)
- Pooh and measure the tubing string out while checking for holes and corrosion pits.
- Pooh and lay down the stator. (The stator is difficult to check for any internal damages.) The stator may be sent back for replacement).
- Lay down, inspect, and replace equipment below the stator assembly if necessary.

Advantages of Progressing Cavity Pumps

- Simple pump design
- Easy-to-manage operation
- Rugged-build parts
- Variable speed control drive
- Will displace heavy fluid, some sand, and solids
- Less energy power requirements

Disadvantages of Progressing Cavity Pumps

- Limited to shallow pumping depths of 4500′
- Holes in the tubing string caused by constant rotational friction
- Rod box failures caused by friction and rubbing on the tubing string

- Will get stuck with sand and solids and create extensive damage to the rotor's and stator's internal parts
- Packing leaks
- May back off tubing
- May twist off sucker rods
- Limited to steel sucker rods only
- Cannot use fiber rod

VII
CHAPTER

PLUNGER LIFT SYSTEM

The plunger lift is used to artificially lift low bottom-hole pressure wells. Plunger lift is a simple and low cost artificial system. It is originally designed to lift and displace accumulated liquid above the perforation in gas wells.

Plunger lift creates mechanical isolation to trap and build high pressure gas pressure to lift off the tool and fluid. Plunger lift uses wellbore gas and/or injection gas pressure to seal off around the plunger, lift fluid and also avoid fluid leak off.

Low cost plunger lift is a good alternative to lift fluid in oil and gas wells with low bottom hole pressure. Plunger lift can be used in the wells with high gas oil ratio fluids that is problematic in other lifting methods such as artificial beam pumping

The principal idea in plunger lift is the fluid mechanical seal below and around the plunger that force the tool up the hole, and allowing gas and oil to surface.

The higher pressure build up below the tool will lift the plunger tool with fluid and force flow the fluid of oil, gas and water to surface. When the high pressure declines or bleed off at the surface; then the plunger will drops off from lubricator at surface to bottom. There are several types of plunger tools that can be used based on wellbore conditions

Plunger lift is one of the simple, efficient, and cost effective artificial lifting system

plunger tool is used to de-liquefy the gas wells in order to maintain the low producing gas wells flowing

The concept of plunger lift tool is to create fluid interface with plunger tool to lift fluid and avoid liquid slippage.

The lifting mechanism of plunger lift consists of a seating nipple as landing point with spring assembly to control pushing, lifting and falling velocity.

The main purpose of plunger tool is to travel up and down to load and unload fluid. On the upstroke, the plunger will be pushed up the hole by the pressure build up under the tool, lifting and displacing fluid to surface. Once the plunger reaches in the surface lubricator, the liquid will be unloaded above the tool, and plunger will fall free down the tubing string to landing point.

There are several styles of plungers used in plunger lifting process based on wellbore fluid conditions such as sand, solids or high gas-liquid ratio fluid

Plungers are made of carbon steel, alloy steel, and/or stainless steel

KHOSROW M. HADIPOUR

INDEX

Printed in the United States
by Baker & Taylor Publisher Services